ARS ELECTRONICA 94

INTELLIGENTE AMBIENTE
intelligent environment

BAND 2

HERAUSGEBER / editors : KARL GERBEL, PETER WEIBEL

REDAKTION / editing : KATHARINA GSÖLLPOINTNER

PVS VERLEGER

ARS ELECTRONICA
Festival für Kunst, Technologie und Gesellschaft
Festival of Art, Technology and Society

Veranstalter/organizers: Brucknerhaus Linz
Vorstandsdirektor Karl Gerbel
Linzer VeranstaltungsgesmbH
Untere Donaulände 7, A-4010 Linz
Telefon: +43/(0)732/7612-0 Fax: +43/(0)732/7612-350
ORF Landesstudio Oberösterreich
Intendant Dr. Hannes Leopoldseder
Europaplatz 3, A-4010 Linz
Telefon: +43/(0)732/6900-0 Fax: +43/(0)732/6900-270

Direktorium/management committee:
Vorstandsdirektor Karl Gerbel (LIVA)
Intendant Dr. Hannes Leopoldseder (ORF)
o.Prof. Peter Weibel (Künstlerischer Leiter)
Ständiger künstlerischer Beirat/
permanent artistic advisory board:
o.Prof. Peter Weibel (artistic director)
Dr. Katharina Gsöllpointner (LIVA)
Dr. Christine Schöpf (ORF)

Herausgeber/editors:
Karl Gerbel/Peter Weibel
Redaktion/editing:
Katharina Gsöllpointner

Übersetzungen/translations:
Tom Appleton, Dipl.-Dolm. Maria E. Clay, David Dempsey, Gender et alia Übersetzungen (Roger Buergel, Susanne Lummerding, Ruth Noack, Georg Tillner), Dr. Waltraud Kolb, Camilla Nielsen, Steve Wilder, B.A., M.A.

Umschlagbild/cover picture/logo design:
Peter Weibel
Computergrafik/computer graphics: Constanze Ruhm
Coverkonzept und Layout/cover-concept and layout:
Loys Egg
Grafik/graphics: Daniel Egg
Satz/type setting: Dr. Vrääth Öhner
Druck/printing: Rema*print*, Wien

Verlag/publishers: PVS Verleger
Vertrieb/sales: PVS Verleger
Pouthongasse 26, A-1150 Wien
Telefon & Fax: +43/1/9832855

Wir danken für die Unterstützung von Ars Electronica:
Stadt Linz, Land Oberösterreich,
Bundesministerium für Unterricht und Kunst

Organisationssekretariat/office: Brigitte Renzl
Produktionsassistenz/production assistant: Karin Sladko
Technische Assistenz/technical assistant: Michael Pointner
Public Relations: Mag. Michaela Kornfehl

ISBN 3-901196-137

VORWORT
preface
KARL GERBEL

Ars Electronica '94 beschäftigt sich mit der Veränderung unserer Umwelt durch den Computer. Die uns bisher bekannte „natürliche" Umwelt wird immer mehr zu einer „künstlichen" Umgebung, die von elektronischen Medien gesteuert wird. Ob Telefon, Fernsehen, Video- und Computerspiele, E-mail, Internet oder Pop-Konzerte, ebenso wie Architektur und Städtebau – der Alltag ist ohne Computer nicht mehr denkbar.

Die Schnittstellen dieser Entwicklung aufzuspüren und dadurch die Welt mit den Mitteln der Kunst mitzugestalten, sie vielleicht auch besser verstehen zu helfen, ist Aufgabe von Ars Electronica.

Ars Electronica 94 hat diesen Tendenzen den Namen „Intelligente Ambiente" gegeben und zum Generalthema des Festivals gemacht.

Ars Electronica 94 bedeutet auch 15 Jahre Ars Electronica in Linz. 1979 wurde dieses Festival für Kunst, Technologie und Gesellschaft begründet und gemeinsam von LIVA/Brucknerhaus und dem ORF/Landesstudio Oberösterreich veranstaltet. Damit war ein neuer Typus von Festival geboren, der in seiner Vernetzung von elektronischer Kunst, Computerkunst, Medienkunst, Performances, elektronischer Festkultur und philosophisch-gesellschaftlichen Perspektiven bis heute beispielgebend ist.

The Ars Electronica 94 deals with the changes to our environment which stem from the computer. The previously familiar "natural" environment is becoming increasingly "artificial" and is falling under the control of electronic media. Telephone, television, video and computer games, e-mail, Internet and pop concerts, as well as architecture and urban planning: Everyday life can no longer be imagined without the computer.

Tracking down the interfaces of this development and thereby helping to shape the world with the help of art, and possibly even understanding it better, is the goal of the Ars Electronica.

The Ars Electronica 94 has given these tendencies the name "Intelligent Ambiences" and made them the theme of the festival.

The Ars Electronica 94 also represents 15 years of the Ars Electronica in Linz. In 1979, this festival for art, technology and society was founded and jointly organized by the LIVA/Brucknerhaus and the ORF provincial studio in Upper Austria. This was the birth of a new kind of festival, and its integration of electronic art, computer art, media art, performances, electronic festival culture and philosophical/societal perspectives continues to serve as an example.

INTELLIGENTE AMBIENTE – CYBER ART

intelligent ambients – cyber art

PETER WEIBEL

Platons Höhlen-Modell der Welt wird neu interpretiert: die Schnittstelle als Vorhang. Tritt ein Beobachter vor die Leinwand, erfaßt ihn eine Videokamera. Die (analogen) Signale der Videokamera von den Bewegungen des Beobachters gelangen zum Computer, wo sie in digitale Zeichenketten verwandelt werden und die digitalen Signalsequenzen der Höhlenmalerei, die im Computer gespeichert sind, beeinflußen. Beide Signalfolgen werden mittels Datenbeamer auf die Leinwand projiziert. Der Beobachter steht physisch vor der Leinwand, virtuell hinter der Leinwand und ist Teil der Felswand bzw. Leinwand. Er ist in den Vorhang eingewebt, in die Höhlenmalerei selbst eingemalt. Er wird virtuell zum externen Beobachter, der von außen an die Höhlenwand drückt.

Plato's cave model of the world is newly reinterpreted: the interface as curtain. If an observer steps up to the canvas, he is caught by a video camera. The analogue signals of the camera, recording the movements of the observer, are passed on to the computer where they are transformed into digital strings of signs, influencing the digital signal sequences of the cave painting stored within the computer. Both rows of signals are projected onto the screen by means of a data beamer. The observer, physically standing before the canvas viz rockface, yet virtually standing behind the canvas, is part of the rockface viz canvas. He is woven into the curtain, painted in within the cave painting itself. He becomes, virtually, an external observer who, from the outside, presses on against the rockface.

Peter Weibel / Bob O'Kane, DER VORHANG VON LASCAUX, interaktive Computerinstallation, 1993

Interactive computer- and video-installations, interactive television, CD-Rom, CDTV, CD-I, laserdisks, musical instrument digital interfaces (MIDI), multimedia PC, virtual reality, hyper media, e-mail, glass fibre cables, video games, video phones, integrated service digital network (ISDN), cyberspace installations, intelligent fassades and buildings – all these interactive image-, text-, and sound-generating model worlds are built upon digital, silicon-based information technology and can be summarily referred to as Cyber Art. They operate with acoustic and visual information as the "new intelligent energy" of the 21st Century and thus cannot deny their origins out of information theory and cybernetics, out of the hystory of the optophonetic arts and kinetics. This exhibition and this book offer a first glimpse of the new Cyber Art direction in the arts and its interrelated original view of the evolution of technology. These artistic model worlds, ranging from cyberspace- and virtual reality-installations to multimedia-networks, demonstrate in an exemplary fashion the basic shift we are experiencing in our world at the end of the 20th Century,

Interaktive Computer- und Video-Installationen, interaktives Fernsehen, CD-Rom, CDTV, CD-I, Laserdisks, Musical Instrument Digital Interfaces (MIDI), Multimedia PC, Virtual Reality, Hypermedien, E-mail, Glasfaserkabel, Videospiele, Videophones, Integrated Service Digital Network (ISDN), Cyberspace Installationen, intelligente Fassaden und Gebäude - all diese interaktiven Bild-, Text-, Ton-Modellwelten beruhen auf der digitalen, siliziumbasierten Informationstechnologie und können als Cyber Art zusammengefaßt werden. Sie arbeiten mit akustischen und visuellen Informationen als die „neue intelligente Energie" des 21. Jahrhunderts und können daher ihren Ursprung aus der Informationstheorie und Kybernetik, aus der Geschichte der optophonetischen Künste und Kinetik nicht verleugnen. Diese Ausstellung und dieses Buch geben einen ersten Einblick in die neue Kunstrichtung Cyber Art und die damit verbundene neue Sicht auf die Evolution der Technologie. Künstlerische Modellwelten, von Cyberspace- und Virtual Reality-Installationen bis zu Multimedia Netzwerken, zeigen beispielhaft eine grundlegende Verländerung unserer Umwelt zu Ende des 20.Jahrhunderts von einer natürlichen , sich selbst überlassenen Umwelt zu einer künstlichen Umwelt, die künstliche Intelligenz besitzt, von einer passiven Umgebung zu einem interaktiven Partner. Die interaktiven Kunstwelten der Cyber Art verweisen auf die zukünftigen intelligenten Ambiente, jene umfassenden, alles vernetzenden, künstlichen, intelligenten Umwelten, die aus den gegenwärtigen digitalen Mensch-Maschine-Medien Schnittstellen, intelligenten Produkt-Ensembles und künstlichen Prothesen hervorgehen werden. Sie zeigen auch die Evolution der Technologie in neuem Licht. Sie zeigen nämlich, daß wir stets Behinderte sind ohne es zu wissen und daß wir die Technologie vorantreiben, um fehlende oder schwache Funktionen der natürlichen Organe zu ersetzen bzw. zu verstärken. Der Behinderte, der mit Hilfe von technischen Prothesen lebt, wird zu einer Modellfigur, zur Avantgarde des Design und der technischen Revolution.

Die Auswahl der Werke zeigt belebte Bildwelten, die auf den Beobachter reagieren, verborgene Informationen preisgeben, aber auch belebte Materialwelten. Wer durch diese Ausstellung geht, deren Design dankenswerterweise von Fareed Armaly ist, hört den Klang der Zukunft. Psychoscapes und Landscapes, Möbellandschaften und Stadtlandschaften verschmelzen, Wände bewegen sich durch die Anwesenheit des Betrachters, die Umgebung ist ein Teil des Organismus, die Objekte haben eine zerfranste Grenzlinie zur Umgebung. Borderline der Objekte, Fuzzy Logik der Organisation, die Präzision des Vagen, und Kovarianz der Subjekte bestimmen die ästhetische Erfahrung. Diese Stimulation des Bewußtseins mit Hilfe künstlicher Sinne und Schnittstellen weichen das Gefängnis der Umwelt auf und machen aus der ästhetischen eine kognitive Erfahrung. Variable Positionen des Subjekts, variable Zonen der Schnittstellen zur Welt, variable Schichten der Visibilität - die Bedingungen des postmodernen Lebens werden überbelichtet und überdeutlich erfahrbar.

from a natural, self-organising environment towards an artificial environment possessed of artificial intelligence, from a passive surrounding towards an interactive partner. The interactive artistic worlds of Cyber Art point up to the intelligent ambients of the future, those all-comprehensive, network-linked, artificial, intelligent environments which will emerge from the present digital Man-machine-media interfaces, from intelligent product ensembles and artificial prostheses.

Peter Weibel, Intelligente Produkte, Prodomo-Wien, 1994

They also throw new light on the evolution of technology, for they show that, unbeknownst to ourselves, we are always handicapped and that we keep pushing onward the developments of technology in order to replace or strengthen missing or weak functions of our natural organs. The handicapped person, living with the aid of technical prostheses, thus becomes a role model, a model figure in the avantgarde of design and of the technological revolution.

The selection of works shows viable pictorial worlds which react with the observer, relinquishing hidden information but also viable material worlds. The observer, moving through this exhibition, whose design, it is gratefully acknowledged, is by Fareed Armaly, hears the sounds of the future. Psychoscapes and landscapes, furniturescapes and cityscapes blend, walls move by dint of the presence of the viewer, the surroundings are a part of the organism, the objects have a fuzzy borderline with their environs. Borderline of the objects, fuzzy logic of the organisation, precision of the vague, and co-variance of the objects determine the aesthetic experience. This stimulation of consciousness with the aid of artificial senses and interfaces dissolves the prison of the environment and transforms the aesthetic experience into a cognitive one. With variable positions of the observer, variable zones of intersection with the world, variable layers of visibility, the conditions of postmodern life become overexposed and can be experienced with an almost hurtful clarity.

Peter Weibel, ALLE TECHNOLOGIE IST FERNTECHNOLOGIE, All technology is tele-technology, Prodomo-Wien, 1994

Peter Weibel, DAS ALPHABET IST NOCH IMMER DIE BESTE KRÜCKE DER WELT, The alphabet – still the best crutch for the world, 1994. SATZBAU – BAUSATZ W1, words into sentences – building blocks, 1988

INHALT
content

KARL GERBEL
VORWORT
preface 3

PETER WEIBEL
INTELLIGENTE AMBIENTE – CYBER ART 4

SYMPOSIEN 8

INTELLIGENTE ENVIRONMENTS

PONTON EUROPEAN MEDIA ART LAB
PIAZZA VIRTUALE: SERVICE AREA A.I. 12

KATHY RAE HUFFMAN / CAROLE ANN KLONARIDES
INTELLIGENT AMBIENCE: VIDEOPROGRAMM 16

MARK TRAYLE
GATES – SEVEN AND HAUNTED 26

CHARLY STEIGER
FLOW 32

KARIN HAZELWANDER
CINEMA SPIKE 34

ROBERT JELINEK
SABOTAGE XIII 36

FÜRST/THALER
ATLANTIS CONSTRUCT 39

ELFRIEDE JELINEK / GOTTFRIED HÜNGSBERG / HANNES FRANZ
TRIGGER YOUR TEXT 40

CONSTANZE RUHM / PETER SANDBICHLER
SECRET OF LIFE 42

ALBA D'URBANO
DER NEGIERTE RAUM
the negated room 46

CYBER ART

FAREED ARMALY
PLAN 50

PERRY HOBERMAN
BAR CODE HOTEL 54

HERMEN MAAT / RON MILTENBURG
SPATIAL LOCATIONS, VERSION III 58

SUPREME PARTICLES
ARCHITEXTURE 64

MONIQUE MULDER / DIRK LÜSEBRINK / GIDEON MAY
E-H 68

MICHAEL KLEIN
CHAOS CUBE – INTERAKTIVE MODELLWELT
chaos cube – interactive model worlds 71

CHRISTIAN MÖLLER
ELECTRONIC MIRROR 2 75

MARTIN KUSCH
AS FAR AS IN-BETWEEN — 78

MICHAEL BIELICKY
DER INTELLIGENTE BRIEFTRÄGER
the intelligent mailman — 82

JEFFEREY SHAW
GOLDEN CALF — 84

WOODY VASULKA
BROTHERHOOD – TABLE III — 87

CHRISTIAN MÖLLER / JOACHIM SAUTER
NETZHAUT — 91

ART + COM
CYBER CITY FLIGHT — 95

ZEITPLAN
schedule — 98

C Y B E R   M U S I C

LOREN & RACHEL CARPENTER
ALLES SPIEL
audience participation — 102

STADTWERSTATT
ELEKTRONISCHES FEST — 106

JARON LANIER
DER SOUND VON EINER HAND
the sound of one hand — 110

ELLIOT SHARP / SOLDIER STRING QUARTET
X-TOPIA — 116

KEN VALITSKY / SOLDIER STRING QUARTET
OILY SAM — 118

GÜNTHER RABL
KARTHASIS
carthasis — 120

CRISTIAN MÖLLER / STEPHEN GALLOWAY
ELECTRO CLIPS — 122

FRIEDRICH GULDA AND HIS PARADISE BAND
FIESTA ELECTRA
Mega-Techno-Rave-into Paradise-Party — 125

UNDERGROUND RESISTANCE / STATION ROSE / UNITED FREQUENCES OF TRANCE
INTELLIGENT/AMBIENT TECHNO — 126

O.Ö. LANDESMUSEUM FRANCISO CAROLINUM
15 JAHRE ARS ELECTRONICA
15 years ars electronica — 129

ORF LANDESSTUDIO
PRIX ARS ELECTRONICA 94 — 133

ÖKS **COMPUTER UND SPIELE**
computer and games — 137

OFFENES KULTURHAUS
MIT DEN AUGEN DER ARCHITEKTUR
with the eyes of architecture — 138

GALERIE MAERZ
PUBLIC INTERVENTION — 141

BIOGRAPHIEN — 142

SYMPOSIEN

INTELLIGENT PRODUCTS
Brucknerhaus, Stiftersaal
**Tuesday, June 21,
10:00 AM – 1:00 PM**

This symposium deals with the emancipation of the tools. The third information-based communications revolution has radically altered the character of the objects. The tool has risen from being an extension of the body to being a machine which processes symbols. The processing of symbols is no longer a privilege reserved to humans; machines are now capable performing this task also. Objects thereby become variables and autonomous agents, materialized chains of symbols that independently process symbols. Intelligent products demonstrate that we will remain handicapped without knowing it. Being handicapped as the avant-garde of design.

Participants:
MANUEL DELANDA
(Doros Motion, New York, USA)
KAY FRIEDRICHS (Universität Karlsruhe, D)
RICH GOLD (Xerox Parc, Palo Alto Research Center, San Francisco, USA)
DAVE WARNER (Loma Linda University Medical Center, CA, USA)

Moderation: PETER WEIBEL

INTELLIGENTE PRODUKTE
Brucknerhaus, Stiftersaal
**Dienstag, 21. Juni,
10.00 – 13.00 Uhr**

Das Symposium bezieht sich auf die Emanzipation der Werkzeuge. Die dritte informationsbasierte Kommunikationsrevolution hat den Charakter der Gegenstände radikal verändert. Vom Werkzeug als Extension des Körpers vollzog sich ein Aufstieg zur symbolverarbeitenden Maschine. Nicht länger ist Symbolverarbeitung ein Privileg des Menschen, neuerdings können das auch Maschinen. Gegenstände werden dadurch zu Variablen und autonomen Agenten, zu materialisierten Zeichenketten, die selbständig Zeichen verarbeiten. Intelligente Produkte zeigen, daß wir stets Behinderte sind, ohne es zu wissen. Behindertsein als Avantgarde des Design.

Teilnehmer:
MANUEL DELANDA
(Doros Motion, New York, USA)
KAY FRIEDRICHS (Universität Karlsruhe, D)
RICH GOLD (Xerox Parc, Palo Alto Research Center, San Francisco, USA)
DAVE WARNER (Loma Linda University Medical Center, CA, USA)

Moderation: PETER WEIBEL

LIFE IN THE NET
Brucknerhaus, Stiftersaal
**Tuesday, June 21,
2:30 PM – 6:00 PM**

Nowadays, virtual electronic worlds are determining our lives as strongly as architectural, urban and rural worlds. We do not live only in houses and streets, but also in cable-channels, telephone- and computer-networks. We are surfing on electronic waves and on water. Bodyless communication on electronic highways prevents traffic jams. Network experts, former hackers, infotrain users, TV specialists and researchers in the field of intelligent products will discuss life in everyday networks.

LEBEN IM NETZ
Brucknerhaus, Stiftersaal
**Dienstag, 21. Juni,
14.30 – 18.00 Uhr**

Virtuelle, elektronische Räume bestimmen mittlerweile unser Leben ebenso wie architektonische, urbane und rurale Räume.
Wir wohnen nicht mehr nur in Häusern und Straßen, sondern auch in Kabelkanälen, Telefonleitungen und Computernetzwerken. Wir surfen auf elektromagnetischen Wellen, nicht nur auf Wasser. Körperlose Kommunikation auf Datenbahnen kennt keinen Stau. Netz-Spezialisten, Ex-Hacker, Infobahn-User, TV-Macher und Forscher aus dem Bereich der intelligenten Produkte diskutieren über neue Modelle des Lebens im vernetzten Alltag.

Teilnehmer:

HANS HÜBNER (ART + COM, Berlin, D)
TOSHIO IWAI (Tokyo, J)
HOWARD RHEINGOLD
(San Francisco, USA)
JEET SINGH
(Art Technology Group, Cambridge, MA, USA)
KEIGO YAMAMOTO
(K-bit Institute, Tokyo, J)

Moderation: JACOB STEUERER

ARCHITEKTUR UND ELEKTRONISCHE MEDIEN

Brucknerhaus, Stiftersaal

**Mittwoch, 22. Juni,
10.00 – 13.00 Uhr und
14.30 – 18.00 Uhr**

Der Einfluß des Computers auf Architektur und Städteplanung hat eine starke Veränderung ästhetischer und kommunikativer Ausdrücke bewirkt. Die Inhalte von Wohnungen und Gebäuden, wie Mikrowellenherd etc. sind oft intelligenter als die Gebäude selbst. Die Autotür sagt mehr über den Zustand des Autos und des Fahrers aus als die Wohnungstür über den Zustand der Wohnung. Flugzeug- und Auto-Technologie werden zu Informationsparadiesen, das Gebäude zur Konsole und die Architektur zur Interface-Technologie.

Über interaktive Architektur, intelligente Gebäude und computergesteuerte elektronische Environments diskutieren Architekten, Baukünstler und Theoretiker.

Teilnehmer:

EDOUARD BANNWART
(ART+ COM, Berlin, D)
ZAHA M. HADID (London, GB)
KEI´ICHI IRIE
(Power Unit Studio, Tokyo, J)
TOYO ITO (Tokyo, J)
CAROLE ANN KLONARIDES
(Long Beach Museum of Art, CA, USA)
SELIM KODER
(Eisenman Architects, USA)
RÜDIGER LAINER (Wien, A)
WOLF D. PRIX (Coop Himmelblau,
Wien – Los Angeles, A)
KEN SAKAMURA (University of Tokyo, J)
MICHAEL SORKIN (New York, USA)

Moderation: KATHARINA GSÖLLPOINTNER

Participants:

HANS HÜBNER (ART + COM, Berlin, D)
TOSHIO IWAI (Tokyo, J)
HOWARD RHEINGOLD
(San Francisco, USA)
JEET SINGH
(Art Technology Group, Cambridge, MA, USA)
KEIGO YAMAMOTO
(K-bit Institute, Tokyo, J)

Moderation: JACOB STEUERER

ARCHITECTURE AND ELECTRONIC MEDIA

Brucknerhaus, Stiftersaal

**Wednesday, June 22,
10:00 AM – 1.00 PM and
2:30 PM – 6.00 PM**

The interrelation of computers, architecture, city planning and electronic environment has lead to a radical change of aesthetic expression and communication patterns. Most buildings are not as intelligent as the machines they contain. A car-door tells me more about the state of the car and my own state than a front-door of the state of the apartment. Airplane- and car-technology are becoming information paradises and architecture becomes a field of interface-technology. Architects, construction artists and theorists will discuss how much architecture and electronics really know about each other.

Participants:

EDOUARD BANNWART
(ART+ COM, Berlin, D)
ZAHA M. HADID (London, GB)
KEI´ICHI IRIE
(Power Unit Studio, Tokyo, J)
TOYO ITO (Tokyo, J)
CAROLE ANN KLONARIDES
(Long Beach Museum of Art, CA, USA)
SELIM KODER
(Eisenman Architects, USA)
RÜDIGER LAINER (Wien, A)
WOLF D. PRIX (Coop Himmelblau,
Wien – Los Angeles, A)
KEN SAKAMURA (University of Tokyo, J)
MICHAEL SORKIN (New York, USA)

Moderation: KATHARINA GSÖLLPOINTNER

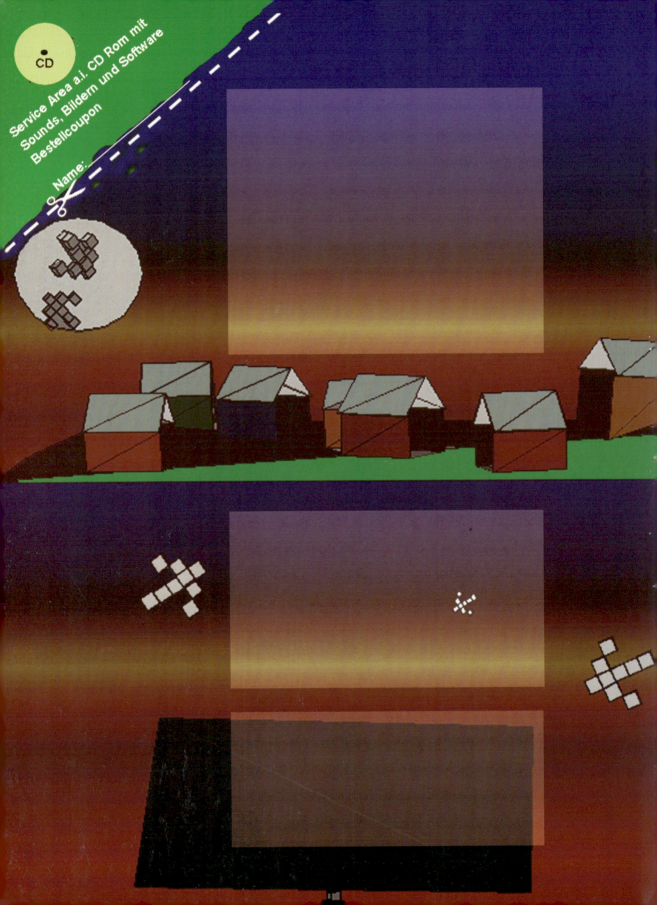

INTELLIGENT AMBIENCE
Video Programm

■ Zusammengestellt von KATHY RAE HUFFMAN und CAROLE ANN KLONARIDES

1. Interim: Within and Beyond Confinement Total: 136:00 min

BILL VIOLA – Reasons for Knocking at an Empty House, 1983, 19:11 min
TOM KALIN – Darling Child, 1993, 1:54 min
SHELLY SILVER – Getting In, 1985, 2:47 min
MICA-TV – The In-Between, 1990, 12:00 min
LESLIE THORNTON & RON VAWTER – Strange Space, 1992, 3:00 min
EDER SANTOS -ESSA COISA NERVOSA (This Nervous Thing), 1991, 15:26 min
JASNA HRIBERNIK – Staircase, 1992, 16:04 min
MICHEL AUDER – Brooding Angels: Made for R.L., 1988, 6:00 min
GARY HILL – Solstice d'Hiver, 1990, 60:00 min

2. Interference: The Invisible Matrix Total: 140:00 min

THERESE SVOBODA – Rogue Transmissions, 1993 1:00 min
JOHN GOFF – Radio Image, 1990, 6:00 min
ERIC M. FREEDMAN – surveiller, a text in two bodies, 1994, 13:00 min
DIANE NERWEN & LES LEVEQUE – GASP, 1993, 13:00 min
PAPER TIGER TV – Staking a Claim in Cyberspace, 1993, 30:00 min
MAX ALMY/TERI YARBROW – Utopia, 1994, 5:00 min
STEINA & WOODY VASULKA – In the Land of the Elevator Girls, 1990, 4:00 min
LES LEVINE – The Media Cage, 1993, 9:00 min
KEN KOBLAND – Stupa, 1992, 60:00 min

3. Interstitial: Between What Is (Seen) Total: 140:00 min

ANNA STEININGER – Going Nowhere Fast, 1993 10:00 min
MICA-TV – Cascade/Vertical Landscapes, 1988 6:30 min
BOB SNYDER – Trim Subdivisions, 1981, 6:00 min
VAN MCELWEE – Inside, 1986, 4:20 min
SHELLY SILVER - Things I Forget To Tell Myself, 1988, 1:50 min
BILL VIOLA – Angel's Gate ,1989, 4.50 min
LARS SPUYBROEK/MAURICE NIO – NOX: Soft City, 1993, 5:00 min
HERMAN VERKERK – Swimmingpool Library, 1993, 25:00 min
SRECO DRAGAN – Arheus (God's Whip), 1992-93, 7:45 min
BETTY SPACKMAN / ANJA WESTERFRÄLKE A I B, 1993, 8:08 min

MICHEL AUDER – Journey to the Center of the Phone Line (a work in progress), 1994, 60:00 min

4. Intervention:
Der taktische Tourist Total: 142:00 min

TONY OURSLER & CONSTANCE DEJONG –
Joyridetm, 1988, 14:23 min
BRANDA MILLER – Time Squared, 1987, 7:00 min
FRANCESC TORRES – Sur del Sur, 1990, 16:24 min
STEINA VASULKA – Urban Episodes, 1980, 8:50 min
NANCY BUCHANAN –
American Dream #7: The Price is Right?, 1991, 13:00 min
EDIN VELEZ – As Is, 1985, 14:05 min
BRUCE YONEMOTO, MELISSA TOTTEN, ED. DE LA TORRE
Banham/Davis Excerpt, 1993, 8:00 min
STEFAAN DECOSTERE – TRAVELOGUE FOUR:
Coming From the Wrong Side, 1992, 56:00 min

■ Intelligent Ambience –
Video Programm

zusammengestellt von Kathy Rae Huffman
und Carole Ann Klonarides

MAX ALMAY & TERI YARBROW
Utopia, 1994, 5 Min.
Utopia arbeitet mit dem Begriff der „Interaktivität"; ein Videospiel, in dem es um das soziale und gesellschaftliche Bewußtsein im heutigen Kalifornien geht. Der Zuschauer/ Spieler wählt scheinbar aus einem Menü utopische oder dystopische Realitäten aus – doch bleibt die Punktezahl unverändert, man gewinnt, wenn man verliert, und umgekehrt.

MICHEL AUDER
Voyage to the Center of the Phone Line,
1991-1994 (Video), 1985-1986 (Audio), 60 Min.
Video-Tagebuch-Schreiber Auder verbrachte seine Sommerferien damit, aus seinem Fenster das endlose Brechen der Meereswellen auf Video aufzunehmen. Dabei ist ihm die Haltung des Voyeurs lieber als die des Weisen; er schafft eine eigene Intimität durch eine bearbeitete Auswahl von Audio-Sequenzen aus vielen Stunden unerlaubt mitgeschnittener Mobiltelefon-Gespräche, die er damit aus der Limousinen-Intimität neurotischer Pendler herauslöst, die auf den „super highways" von Long Island, New York, ins Nirgendwo rasen.

Brooding Angels: Made for R.L., 1988, 6 Min.
Eine persönliche Sicht des urbanen Lebens als Zusammenschnitt von Material, das Auder aus dem Fenster seiner Wohnung im 20. Stockwerk in Manhattan und vom Fernsehapparat in der Wohnung aufgenommen hat. Durch die Konstrastie-

MICHEL AUDER – Journey to the Center of the Phone Line (a work in progress), 1994, 60:00 min

4. Intervention:
The Tactical Tourist Total: 142:00 min

TONY OURSLER & CONSTANCE DEJONG –
Joyride™, 1988, 14:23 min
BRANDA MILLER – Time Squared, 1987, 7:00 min
FRANCESC TORRES – Sur del Sur, 1990, 16:24 min
STEINA VASULKA – Urban Episodes, 1980, 8:50 min
NANCY BUCHANAN – American Dream #7: The Price is Right?, 1991, 13:00 min
EDIN VELEZ – As Is, 1985, 14:05 min
BRUCE YONEMOTO, MELISSA TOTTEN, ED. DE LA TORRE Banham/Davis Excerpt, 1993, 8:00 min
STEFAAN DECOSTERE – TRAVELOGUE FOUR: Coming From the Wrong Side, 1992, 56:00 min

Intelligent Ambience –
The Video Theater Program

curated by Kathy Rae Huffman
and Carole Ann Klonarides

MAX ALMAY & TERI YARBROW
Utopia, 1994, 5 min
Playing off of the notion of "interactive", Utopia poses itself as a video game that is plugged into the social consciousness of contemporary California. The viewer/player seemingly makes choices from the menu offering utopian or dystopian realities – however, the score is always the same; you win while you lose, and vice-versa.

MICHEL AUDER
Voyage to the Center of the Phone Line, 1991-94 (video), 1985-86 (audio) 60:00 min
Video Diarist Auder spent his summer vacation videotaping out his window the endless break of the ocean. Preferring the stance of a voyeur to a sage, Auder creates his own intimacy through his choice of audio excerpts edited from hours of pirated cell phone conversations, technically lifted from the limo privacy of neurotic commuters speeding towards nowhere on the "super highways" of Long Island, New York.

Brooding Angels: Made for R.L., 1988, 6 min
An edited personal vision of urban life collected from material shot from the window of his twentieth floor Manhattan studio and off the television monitor contained within. His contrast of mediated/ experienced material heightens the meaning of both. A grotesquerie of urban life

ensues with an occasional idyllic vision of nature.

NANCY BUCHANAN
American Dream #7: The Price is Right, 1991, 13:00 min

Home ownership is an American dream and considered a basic right. But, the low interest loans and affordable property that were a reward to WWII veterans, is now a fantasy. The political control over the utopian living environment of Los Angeles, a metropolitan landscape without end, has been exposed – and its self-segregated communities revealed as a plan to prevent racial integration and to maintain the wealthy, upper class residential enclaves on the Westside. Mike Davis, author of *City of Quartz*, contributes biting commentary and historic information on California's suburban expansion, and the current battles to provide ample growth for future generations.

STEFAAN DECOSTERE
Travelogue Four:
Coming from the Wrong Side, 1992, 56:00 min

Travelogue is an ongoing investigation, and part four is a mediated "package" that exposes the diversity of place, time and event. It presents The Banff Springs Hotel, a turn of the century luxury hotel in the beautiful Canadian Rockies. A perfectly preserved tourist site in the modern world of advertising and theme parks, the hotel is compelled to compete: Native American dancers perform for the camera and the bystanders, becoming the future – instantly archived.
Production: The Banff Center for the Arts and BRT, for the Belgium Television series BRTN. Composer: John Oswald

SRECO DRAGAN
Arheus (God's Whip), 1992-93, 7:45 min

The artist considers the Renaissance painting "Prospettiva di citta" by Piero della Francesca and the virtual reality of the figure of Arheus. Both are beyond the frame of reference to the experience of the observer. The sound of Laibach, with their version of Italian Renaissance music, further constructs the fiction of space. Production: TV Slovenija. Music: Laibach.

ERIC M. FREEDMAN
surveiller. a text in two bodies, 1994, 13 min

A self-reflexive look at the power dynamics of interpersonal relationships, as mediated through theo-

rung von Medienmaterial und erlebtem Material wird die Bedeutung beider verstärkt. Eine Groteske des urbanen Lebens mit einem gelegentlichen idyllischen Blick in die Natur.

NANCY BUCHANAN
American Dream #7: The Price is Right, 1991, 13 Min.

Ein eigenes Haus zu besitzen gehört zum amerikanischen Traum und gilt als Grundrecht. Die niedrig verzinsten Kredite und erschwinglichen Grundstücke, mit denen man Veteranen des Zweiten Weltkriegs belohnte, sind heute Illusion. Die politische Kontrolle über den utopischen Lebensraum Los Angeles, eine nicht endende Stadtlandschaft, wurde aufgedeckt – und die Abkapselung einiger Bezirke als Plan zur Verhinderung von Rassenintegration und Erhaltung der Wohn-Enklaven der Reichen auf der Westside entlarvt. Mike Davis, der Autor von *City of Quartz*, liefert scharfe Kommentare und geschichtliche Informationen über die suburbane Expansion in Kalifornien und das Bemühen, für zukünftige Generationen ein hohes Wachstum zu garantieren.

STEFAAN DECOSTERE
Travelogue Four:
Coming from the Wrong Side, 1992, 56 Min.

Travelogue ist eine noch laufende Untersuchung; Teil vier ist ein mediatisiertes „Paket", das die Diversität von Ort, Zeit und Ereignis vorführt. Es zeigt das Banff-Springs-Hotel, ein Luxushotel aus der Zeit der der Jahrhundertwende in den kanadischen Rockies. Das Hotel ist ein perfekt konservierter Touristentreff inmitten der modernen Welt der Werbung und der Theme Parks und muß gegen diese Konkurrenz antreten: amerikanische Ureinwohner tanzen vor der Kamera und für die Schaulustigen, werden so zur Zukunft – und sofort archiviert.
Produktion: The Banff Center for the Arts und BRT, für die belgische Fernsehserie BRTN. Komponist: John Oswald

SRECO DRAGAN
Arheus (God's Whip), 1992-1993, 7:45 Min.

Der Künstler betrachtet das Renaissance-Gemälde „Prospettiva di città" von Piero della Francesca und die virtuelle Realität der Figur des Arheus. Beide liegen außerhalb des Bezugsrahmens der Erfahrung des Betrachters. Der Klang von „Laibach", mit ihrer Version italienischer Renaissance-Musik, trägt noch weiter zur räumlichen Fiktion bei. Produktion: TV Slovenija. Musik: Laibach.

ERIC M. FREEDMAN
surveiller. a text in two bodies, 1994, 13 Min.

Ein selbstreflektierender Blick auf die Machtdynamik zwischenmenschlicher Beziehungen, wie sie im theoretischen

retical discourse, telecommunications, and the lens of a video camera.
text: Michel Foucault, "Discipline and Punish", Vintage Books, NY l979

JOHN GOFF
Radio Image, 1990 6:00 min

Electromagnetic rays are visualized with the help of a cathode ray tube and a computer: the radio-image laboriously evolves, attempting to display its heterogeneous sources. A suspended image emerges, and for a brief moment reveals a woman who appears to be delivering the evening news before it fades back into its invisible signal.

GARY HILL
Solstice d'Hiver, l990, 60:00 min

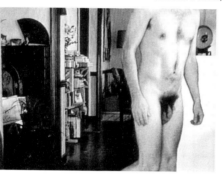

This is Gary Hill's last single channel videotape to date and was shot in real time on December 21, 1990 between the hours of 1:30 p.m. and 2:30 p.m. It begins with the camera moving in almost silent, slow increments around a room, occasionally interrupted by the sound of the autofocus on the camera lens readjusting as it searches for a subject to focus on. A figure (Hill) enters into the frame and with determined slowness places a record on an obsolete technology – a record player. The record continues to play a repeated message while Hill takes a shower, leaving the camera to document. The recording, the camera, and Hill become separate entities.
Commissioned by La Sept for "Live", a series of real time videotapes proposed by Phillipe Grandieux. Sound recording played -Alvin Lucier, "I am Sitting in a Room".

JASNA HRIBERNIK
Staircase, 1992, 16:04 min

The city streets are hectic, but inside an empty factory a psychological post-war narrative commences with the meeting of a man and a woman on a staircase. Against the measure of time a digital stopwatch intrudes, counting back and forth to zero. A struggle ensues, are they imprisoned? The climax reveals their bloody memories of recent history and the Balkan War.
Production: V.P.K., TV Slovenija, Slovenian Ministry of Culture, Studio Ljubljana.
Music: Dome, This Heat

TOM KALIN
Darling Child, 1993, 1:44 min
A visual poem that functions as an alternative music video in response to issues of sexuality and human interaction. Spare visuals of an interior space, and minimalist music create an evocative, metaphoric interpretation of a text by Truman Capote, in this 1990s response to the AIDS crisis.
Music: Brian Eno,

KEN KOBLAND
Stupa, 1992, 60:00 min
A flight over the suburbs yields a relentless observation of the matrix of the communities below. Highways, shopping malls, parks and the endless look-alike rooftops of residences become ambiguous references to the vast burial grounds situated nearby (but out of sight of the living). Throughout the journey, the scanning of radio programs exposes connections between households and commuters. The consciousness of the American public is revealed in talk shows, country western music, and traffic reports – in a never ending soundtrack: a solemn memorial to suburban life and death.

LES LEVINE
The Media Cage, 1993, 9:00 min
Imposing a computer generated "cage" on headshots of individuals speaking in generic sound bites, Levine underscores the feeling of xenophonia created by the invisible architectures which exist around us.

VAN MCELWEE
Inside, I986, 4:40 min
This piece occupies a place where dreams and architecture overlap. A large, odd-angled mall is extended into an endless tunnel, which is then revealed to be a unit in a honeycomb of such spaces.

MICA-TV
(Michael Owen and Carole Ann Klonarides)
The In-Between, 1990, 11:41 min
The Gothic theme of the doppelganger is played out against the sleek modernity of the Wexner Center for Visual Arts at the Ohio State University, designed by Peter Eisenman. Adapted from "Analogue," a short story by Susan Daitch, the fragmented narrative parallels Eisenman's architectural approach, which deconstructs symbolic associations.
Commissioned by the Wexner Center for the Visual Arts at Ohio State university in Columbus, Ohio and the BBC2. Music composed and performed by David Weinstein with Shelley Hirsch.

TOM KALIN
Darling Child, 1993, 1:44 Min.
Ein visuelles Gedicht in der Funktion eines alternativen Musikvideos als Antwort auf Fragen der Sexualität und der zwischenmenschlichen Beziehungen. Sparsame Ansichten eines Innenraumes und minimalistische Musik schaffen eine evokative, metaphorische Interpretation eines Textes von Truman Capote; eine Antwort der 90er Jahre auf die AIDS-Krise.
Musik: Brian Eno

KEN KOBLAND
Stupa, 1992, 60 Min.
Ein Flug über Vorstadtviertel liefert ein unbarmherziges Bild der Matrix dieser Siedlungen. Autobahnen, Einkaufszentren, Parkanlagen und die endlosen, immer gleich aussehenden Dächer der Wohnhäuser werden zu mehrdeutigen Verweisen auf die riesigen Friedhofsanlagen in der Nähe (jedoch außer Sichtweite für die Lebenden). Durch das ständige Hin- und Herschalten zwischen Radioprogrammen während des Flugs werden Verbindungen zwischen Haushalten und Pendlern aufgedeckt. Das Bewußtsein der amerikanischen Bevölkerung offenbart sich in Talk-Shows, Country music und Verkehrsmeldungen – in einem endlosen Soundtrack: ein feierliches Ehrenmal für das Vorstadtleben und den Vorstadttod.

LES LEVINE
The Media Cage, 1994, 9 Min.
Indem er über Menschen, die in Soundbites sprechen, ein von einem Computer gezeichnetes Bild eines Käfigs stülpt, betont Levine das Gefühl der Xenophonie, ausgelöst durch die rund um uns existierende unsichtbare Architektur.

VAN MCELWEE
Inside, 1986, 4:40 Min.
In diesem Stück überschneiden sich Traum und Architektur. Ein großes Einkaufszentrum wird zu einem endlosen Tunnel auseinandergezogen, aus dem dann eine Zelle in einer aus solchen Räumen bestehenden Wabe wird.

MICA-TV
(Michael Owen und Carole Ann Klonarides)
The In-Between, 1990, 11:41 Min.
Das romantische Doppelgängermotiv wird gegen die glatte Modernität des Wexner-Center-for-Visual-Arts der Ohio-State-University, entworfen von Peter Eisenman, ausgespielt. Narrative Fragmente aus einer Bearbeitung der Short Story „Analogue" von Susan Daitch bilden eine Parallele zu Eisenmans architektonischem Konzept, wodurch symbolische Assoziationen dekonstruiert werden.
Im Auftrag des Wexner-Center-for-Visual-Arts der Ohio-State-University in Columbus, Ohio, und BBC 2. Musik von David Weinstein und Shelley Hirsch.

Cascade/Vertical Landscapes, 1988, 6:30 Min.

Die ununterbrochene Abfolge vertikaler Kamerabewegungen und Bildüberblendungen simulieren die Art und Weise, wie wir heute die amerikanische Landschaft erleben. Die Aufnahmen stammen aus Südkalifornien und dem Großraum New York und betonen die amerikanische Architektur, die scheinbar eher dazu gebaut wurde, um fotografiert zu werden, als um bewohnt zu werden. In Zusammenarbeit mit Dike Blair und Dan Graham, Musik von Christian Marclay. Im Auftrag von Channel 4, U.K.

Cascade/Vertical Landscapes, 1988, 6:30 min

Constructed as a continuous parade of vertical camera movements and image layering to simulate the way we experience the contemporary American landscape. The work is shot on locations in Southern California and in the New York City area, highlighting American architecture, which seems built to be photographed rather than inhabited. In collaboration with artists Dike Blair, Dan Graham, and music by Christian Marclay. Commissioned by UK's Channel 4.

BRANDA MILLER
Time Squared, 1987, 7 Min.

Jahrzehntelang war der Times Square einer der markantesten Punkte der Welt. Als im Zuge der Stadterneuerung eine „Verbesserung" der Gegend drohte, protestierten die Bewohner New Yorks entrüstet – und stoppten letztlich den „Fortschritt". In der Arbeit verschmelzen, ähnlich wie in Musik-Videos, Bilder und Klänge der Vergangenheit und der Gegenwart in transparenten Überblendungen und laufen im Büro von John-Burgee-Architects zusammen, wo sich das sterile Modell des zukünftigen Stadtzentrums befindet, frei von Sexualität, Obdachlosen und Straßenleben.
Produktion: The Contemporary Art Television Fund für TIME CODE. Musik: A. Leroy.

BRANDA MILLER
Time Squared, 1987, 7:00 min

For decades, Times Square has been one of the most recognizable spots in the world. When threatened with urban renewal to "improve" the area, New Yorkers were outraged and objected – ultimately stopping "progress." In a music-video style work, images and sounds of the past and present merge – in a transparent overlay and converge in the office of John Burgee Architects, where the sterile model of the future city center is situated, devoid of sex, the homeless and street action.
Production: The Contemporary Art Television Fund for TIME CODE. Music: A. Leroy.

DIANE NERWEN,
in Zusammenarbeit mit LES LEVEQUE
GASP, 1993, 13 Min.

GASP findet in dem Moment statt, in dem die „spektakuläre" Sicht von der Spitze einer „smart bomb" sich in die korporative Rhetorik eines „one world network" auflöst. GASP versucht fallende geographische Grenzen darzustellen, einen virtuellen Raum, in dem Sehen das Innere mit dem Äußeren verschmilzt, das Öffentliche mit dem Privaten und den Körper mit der Maschine. In GASP löst sich die Identität des einzelnen Subjekts (des Reisenden, des Patienten) auf und verschwindet in der rasenden und weltlichen Vereinnahmung elektronischer Simulationen.

DIANE NERWEN,
in collaboration with LES LEVEQUE
GASP, 1993, 13 min

GASP is set in the moment when the "spectacular" view from the nose of a "smart bomb" fades into the corporate rhetoric of a "one world network". GASP attempts to re-present collapsing boundaries of geography, a virtual space where looking merges the internal with the external, the public with the private and the body with the machine. In GASP the identity of the individual subject (the traveler, the patient) is diffused and disappears into the delirious and mundane reception of electronic simulations.

TONY OURSLER und CONSTANCE DEJONG
Joyride (TM), 1988, 14:23 Min.

In einer traumartigen Achterbahnfahrt durch einen Theme Park entfaltet sich das odysseehafte Spektakel der Konsumkultur und des amerikanischen Marktes. In den Augen der

TONY OURSLER and CONSTANCE DEJONG
Joyride (TM), 1988, 14:23 min

In a dreamlike roller-coster ride through a corporate theme park, an odyssey of the spectacle of consumer culture and the American marketplace is unfolded. The artists write that it is "inspired by the institutional versus the private-sector devotion to the transcendental."

A Western front Video Production in association with Los Angeles Contemporary Exhibitions (LACE).

PAPER TIGER TELEVISION
Staking a Claim in Cyberspace, 1993, 30 min

A welcome alternative look at the coming digital highway and the mass media's vision of the electronic consumer. Comparing the high commercial to the grass roots end, this program suggests that the public should take a more active role in developing these new "architectures".

Co-produced by Michael Eisenmenger, Linda Iannacone, Mary Feaster and Cathy Scott

EDER SANTOS
Essa Coisa Nervosa (This Nervous Thing), 1991, 15:26 min

Questioning the ways we perceive and receive information, Santos writes, "lost in our creations we must use artificial means, such as newspapers and other media to simulate knowledge of what is around us. In doing so, we create heroes, cities, characters, icons and monuments." A driving soundtrack propels the viewer through a frenetic trans-cultural landscape.

SHELLY SILVER
Things I Forget To Tell Myself, 1988, 1:54 min

Living in a city one sees parts, rarely the whole. The media environment has extended this condition of anomie. This tape is made of tiny snippets of meaning, shot in a day in New York City, then cut together to form a rhythmic flow of unconscious associations. The giving and withholding of certain information, what we see and don't see, is not often offered as a privilege of choice.

Getting In, 1989, 2:47 min

"Getting in" is about heterosexual sex and the architecture of Northern California. By combining the two, Silver adds a new twist to the idea of a threshold as we continually approach and finally "come" out of buildings. The soundtrack was obtained from a sound recordist's out-takes from a porno film before the music is mixed-in.

BOB SNYDER
Trim Subdivisions, 1981, 6:00 min, (silent)

In a game-like fashion, images of suburban homes are layered and wiped -back and forth between two images- to express the two-dimensionality and architectural redundancy of tract houses in provincial neighborhoods. A sophisticated exploration of the formal properties of image, the digital video effects allow the homes to be seen as a set of interchangeable units – a sameness which pro-

den die Häuser zu austauschbaren Einheiten – eine Gleichheit, die die Vereinheitlichung von Lebensstil und Lebensstandard fördert.

motes the concept of unification in life styles and standards of living.

BETTY SPACKMAN & ANJA WESTERFRÖLKE
A I B, 1993, 8:08 Min.
Die Konzepte Territorium und Kommunikation werden anhand von wiederholten Anweisungen untersucht, die zwei Frauen einander auf dem Wasser erteilen. Treiben, navigieren, den Kurs halten – die Sprache ist der Schlüssel zum Verständnis von Raum und Position. Die Folge sind Frustrationen, und die Notwendigkeit von Grenzen wird erkennbar. Von oben gesehen wird der Fluß zu einem sichtbaren Element im geistigen Aufeinandertreffen von menschlichem Ehrgeiz und der Energie der Natur.
Produktion: Offenes Kulturhaus Linz.
Musik: Thomas Nättling

BETTY SPACKMAN & ANJA WESTERFRÖLKE
A I B, 1993, 8:08 min
An investigation of territory and communication, through the repeated instructions of two women - who instruct each other- on the water. Floating, navigating, and keeping in line – language is the key to understanding space and position. Frustrations ensue and the necessity of border(s) is observed. From an overhead vantage point, the view of the river becomes the visible element in the spiritual encounter between human ambition and nature's energy.
Production: Offenes Kulturhaus Linz.
Music: Thomas Nättling

LARS SPUYBROEK/MAURICE NIO
NOX: Soft City, 1993, 5 Min.
Das NOX-Statement zur Architektur lautet: „In einer Welt ständiger Kontamination mit Theorien, Bildern und Disziplinen sollte die Architektur alle technologischen Möglichkeiten über ihren Nutzen und ihre Richtlinien hinaus parasitieren (sic)." NOX demonstriert anhand der Exzesse der Modernität seine Sicht der Beziehung zwischen moderner Kultur und neuen Planungsansätzen.

LARS SPUYBROEK/MAURICE NIO
NOX: Soft City, 1993, 5:00 min
The NOX – Architecture statement is, "in a world of constant contamination of theories, images and disciplines, architecture should parasitize [sic] all possibilities of technology, beyond the reach of their good use and guiding principles." NOX employs the excesses of modernity to demonstrate their vision of the relationship between modern culture and new design methodology.

ANNA STEININGER
Going Nowhere Fast, 1993, 10 Min.
Eine theoretische Suche nach Möglichkeiten post-ethischer Ausdrucksmittel. Das fast unsichtbare Subjekt wird von Schichten von Bildern umgeben und überwältigt, und keine Abhilfe ist möglich.
Produktion: Medienwerkstatt Wien

ANNA STEININGER
Going Nowhere Fast, 1993, 10:00 min
A theoretical image search towards possibilities for a post-ethical means of expression. The nearly invisible subject is surrounded and overwhelmed with layers of images, from which no relief is possible.
Production: Medienwerkstatt Wien

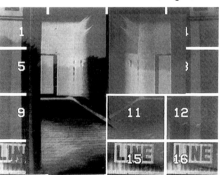

LESLIE THORNTON UND RON VAWTER
Strange Space, 1992, 3 Min.
Während Vawter laut aus einem Gedicht von Rainer Maria Rilke vorliest, hört man, wie ein Arzt über seinen Gesundheitszustand spricht. Fotografische Bilder von inneren Organen und von der Oberfläche des Mondes kontrastieren Landschaften des innersten Raumes und des Weltraumes. Das Ergebnis ist ein quälendes Grübeln über das Verhältnis zwischen und die Disparität von medizinischer Interpretation und persönlichem Erleben der Körperlichkeit, der Sterblichkeit und letztlich der Geistigkeit.
Produziert für den AIDS Awareness Day 1992

LESLIE THORNTON AND RON VAWTER
Strange Space, 1992, 3:00 min
While Vawter reads aloud from a poem by Rainer Maria Rilke, a doctor is heard discussing his medical condition. Photographic images of internal organs and of the moon's surface create landscapes of inner and outer space. The result is a haunting rumination on the relation and disparity between medical interpretations and personal experiences of physicality, mortality, corporeality, and, ultimately, spirituality.
Produced for AIDS Awareness Day 1992

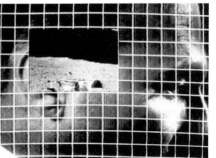

FRANCESC TORRES
Sur del Sur, 1990, 16:24 min
An investigation of the history of Seville, one of the few cities in Spain where one can actually perceive all the layers of its history in a synchronic way. This work explores the coexistence of two complimentary and opposite concepts of time, historical time - which is linear- and mythical time -which is circular.

BRUCE YONEMOTO, MELISSA TOTTEN, ED. DE LA TORRE
Banham/Davis Excerpt: The Architecture of Four Ecologies (Reyner Banham) (City of Quartz (Mike Davis)), 1993, 8 min
A textural, overtly styled music video that juxtaposes two generational views of the California landscape: on the lifestyles of Los Angelenos Banham states, "the freeway is not a limbo of existential angst, but the place where they can spend the two calmest and most rewarding hours of their daily life"; and to update Davis laments, "in the erstwhile world capital of teenagers, where millions overseas still imagine Gidget at a late-night surf party, the beaches are now closed at dark, patrolled by helicopter gunships and police dune buggies."
Commissioned by KCET's *The Works II: The '60s in the '90s*, broadcast June 1993 in Los Angeles.
Music: Tom Recchion

STEINA VASULKA
Urban Episodes, 1980, 8:50 min
Downtown Minneapolis is observed through mechanized camera controls. Ordinarily, the camera's point of view is associated with the human viewpoint, and pays attention to the human condition around it. Here, the mechanism (Machine Vision) directs the observations from a machine's perspective instead – giving new insights to the city and its inhabitants.
Producer: KTCA-TV, Minneapolis,
Optical instrumentation: Josef Frames

STEINA and WOODY VASULKA
In the Land of the Elevator Girls, 1989, 4:00 min
The elevator is the metaphorical vehicle used to reveal an outsider's gaze into contemporary Japanese culture. The continual opening and closing of elevator doors serves as a succinct formal device, as the viewer is offered brief glimpses of a series of landscapes – natural, urban, cultural and domestic. Doors open onto doors to reveal layers of public and private vision, transporting the viewer from theatrical performances and street scenes to an elevator surveillance camera's recording of everyday life.
Producer: IMATCO/ATANOR for Television Espanola S.A. El Arte del Video

EDIN VELEZ
As Is, 1984, 11:28 min
A mythical interpretation of New York City, from the grand theater of its urban architecture to its diverse ethnic heritage. Velez reshapes and layers the urban landscape, isolating gestures and rituals as

paradigmatic signs and symbols within the vast, indifferent metropolis.

HERMAN VERKERK
Swimmingpool Library, 1993, 25:00 min

In the architectural domain, pragmatic perspective usually dominates. This vision exists in a virtual world, and only intermittently interrupts the tangible landscape where it might be built. What began as an attempt to understand the cause and effect of a street that was permanently jammed with traffic, evolved into a hybrid building imagined as a mind machine: a combination swimming pool and library (both excellent thinking places).
Producer: ETH, Zurich, für Nachdiplomstudium Architektur & C.A.A.D.

BILL VIOLA
Angel's Gate, 1989, 4:50 min
Viola writes, "..A succession of individual images focusing on mortality, decay and disintegration... appear as a series of openings or momentary glimpses into nature's essential gestures which, like thoughts, are destined to fade and themselves disintegrate into obscurity and oblivion." The journey of the camera reveals universal symbols, that "pass through and out into the bright world, liberated by the consuming, saturated white light of its own overexposure."

Reasons for Knocking at an Empty House, 1983, 19:11 min

A powerfully austere observation of the perceptual experience of the self in isolation, subjected to extended duration. Viola writes that this work is "an attempt to stay awake continuously for three days while confined to an upstairs room in an empty house. The space becomes increasingly subjective as events slide in and out of conscious awareness and the duration becomes more and more brutal."
Produced in association with the TV Lab at WNET/Thirteen, New York

GATES – SEVEN AND HAUNTED

MARK TRAYLE

I haven't put pen to manuscript paper in years. Since the late 1970s I've been using electronic hardware and computer software to fill in the musical details while I've focused my attention on creating larger musical processes and situations. It's not that I'm lazy, it's just that I bought into the Cageian notion of removing ego from the process of creating music a long time ago, and I can't seem to give it up. I also have an affinity for found sound material...for the gritty distortion and electronic artifacts of radio and television broadcasts, the over-amplified flotsam of popular culture.

Fortunate Circumstances

Ever since I graduated from Mills College in 1982, I've made my living programming computers. Thanks to computer industry folk wisdom that musicians make good programmers, and a basic understanding of digital logic design and programming gained as a byproduct of practicing electroacoustic music, I was able to get a "straight job" after leaving Mills. Since then I've had a succession of such jobs, most of them at least a little interesting in their own right, but not what I'd call musically inspiring. Through a long chain of fortunate circumstances I ended up working at the NASA Ames Research Center in 1987, in one of the first virtual reality labs. I learned enough about VR there to know that a dataglove might make a good performance interface, and that a computer-generated 3D space might be a fertile place to put those Cageian musical processes into action.

Of course at that time VR technology was (and for the most part still is) economically out of my reach. Sometime in the late 1980s Mattel, the American toy company, introduced an inexpensive dataglove for Nintendo video games called the Power Glove. Like the much more expensive datagloves used for "serious" VR research the Power Glove operates within a finite three dimensional field and sends information to a host computer about its location within the field and which of its fingers are bent. The Power Glove also has a set of buttons located on the wrist which send ASCII codes to the host. The low cost (originally about $80, now about $20 if you can find them)

sofern man sie überhaupt noch bekommt) und so weit verbreitet, daß es, so dachte ich, bald in jedem Haushalt einen geben würde, der an irgendeinen nachgerüsteten Spielcomputer angeschlossen ist – das Walzenklavier des 21. Jahrhunderts. Das brachte mich auf die Idee, ein paar Stücke für den Glove zu schreiben, die auch Amateurmusiker spielen könnten, wobei das Interface so einfach zu verwenden war, daß man nicht erst jahrelang üben mußte, um seinen Freunden nach dem Abendessen etwas vorzuspielen. Virtuelle Kammermusik.

Mitte 1990 begann ich, Musiksoftware für den Glove zu schreiben. Kurz zuvor hatte ich bei einer ACM SIGCHI-Konferenz ein Demo gesehen und fand dann auf dem „grauen Markt" auch ein Produkt, durch das ich den Glove mit meinem Amiga verbinden konnte. Die Software, die ich entwickelte (sie ist in Lattice C unter AmigaDOS geschrieben), steuert durch Bewegungen des Glove die Tonhöhe, Amplitude und Länge gesampelter Sounds. Da ich ein erschwingliches Heimgerät entwickeln wollte, verwendete ich das interne Audiosystem des Amiga als Sampler. Obwohl es ein 8-Bit-System mit Low-Fidelity war, konnte ich auf Grund seiner offenen Architektur mit den Samples viel machen, z.B. die Anfangs- und Endpunkte in Echtzeit verändern, was man damals mit kommerziellen Samplern, auch wenn sie eine höhere Wiedergabetreue hatten, nicht konnte.

Gates

Eines der ersten Stücke, das ich für den Glove schrieb, hieß Gate. In dem Stück bzw. in den Versionen des Stückes, die hier beschrieben sind, stellen Bewegungen auf der horizontalen Ebene (x) Veränderungen der Tonhöhe dar, Bewegungen auf der vertikalen Ebene (y) Veränderungen der Amplitude und Bewegungen auf der senkrecht auf x und y stehenden Ebene (z) die Länge der Samples. (Hier ist anzumerken, daß Veränderungen der Samplelänge, zumindest in meiner Musik, normalerweise als Veränderungen der Klangfarbe wahrgenommen werden, wenn die Samples relativ kurz sind, d.h. etwa 0,25 Sekunden oder kürzer, und als Veränderungen des Rhythmus, wenn sie länger sind.) Innerhalb des Feldes können Ebenen definiert und als Auslöser sensibilisiert werden – bewegt man den Glove durch eine dieser imaginären Ebenen, wird ein Ereignis oder eine Ereigniskette ausgelöst. Ich bezeichne diese Ebenen daher oft auch als Auslöser-Ebenen. Die für Gate definierte Auslöser-Ebene verläuft senkrecht zur z-Achse an deren Mittelpunkt und erstreckt sich über die gesamte Länge der x- und y-Achsen. Diese Ebene ist das „Gate", das Tor. Wenn der Glove nach vorn (vom Körper weg) durch die Ebene bewegt wird, „öffnet" sich das Gate, wenn er nach hinten (zum Körper hin) bewegt wird, „schließt" sich das Gate. Immer wenn das Gate geöffnet oder geschlossen wird, wird ein Sample eines sich öffnenden oder sich

and wide distribution of the Power Glove seemed to suggest that there would soon be one in every home, hooked up to a supercharged entertainment computer of some sort... a player-piano for the 21st century. That inspired me to write some pieces for the glove that could be played by amateur musicians, with a performance interface simple enough to use that you wouldn't have to study the instrument for years just to play music for your friends after dinner. Virtual chamber music.

I started writing music software for the glove in mid-1990, shortly after seeing a demo at the ACM SIGCHI conference that year, and finding a "gray-market" breakout box that would allow me to interface the glove to my Amiga. The software that I developed (written in Lattice C under AmigaDOS) uses movements of the glove within the field to control the pitch, amplitude, and duration of sampled sounds. And in keeping with the idea of making an affordable music appliance for the home, I used the Amiga's internal audio system as my sampler. Although it's a low-fidelity 8-bit system its open architecture allowed me to perform tricks with samples, such as changing start and end points in real-time, that you couldn't do with higher-fidelity commercial samplers at that time.

Gates

One of the first pieces I wrote for the glove was called Gate. In this piece and the versions of it described below, motion on the horizontal (x) plane is mapped to changes in pitch, along the vertical (y) plane to changes in amplitude, and along the plane perpendicular to both x and y (z), sample length. (It's worth noting here that changes in sample length are generally perceived, in my music anyway, as changes in timbre when the samples are relatively short, about 0.25 seconds or less, and as changes in rhythm when they are longer). Planes within the field can be defined and made edge-sensitive – passing the glove through one of these imaginary planes will trigger an event or chain of events. I sometimes refer to these planes as trigger planes. The trigger plane defined for Gate runs perpendicular to the z axis at it's midpoint, for the full extents of the x and y axes. This plane is the 'gate'. When the glove moves forward (away from the body) through the plane, the gate 'opens'; when the glove moves back towards the body, the gate 'closes". Whenever the gate is opened or closed, a sample of a gate or door opening or closing is played. On one side of the trigger plane is a collection of audio samples; I call it 'sample space'. On the other side of the trigger plane is an area where the audio samples from sample space can be played, called play 'space'. (see below). The buttons on the glove are used to start, stop or pause the performance, or to load new samples.

schließenden Gates oder Tores gespielt. Auf der einen Seite der Auslöser-Ebene befindet sich eine Sammlung von Audiosamples, ich bezeichne das als den „Sample-Raum". Auf der anderen Seite der Auslöser-Ebene befindet sich ein Raum, in dem die Audiosamples aus dem Sample-Raum gespielt werden können, der sogenannte „Spiel-Raum" (siehe weiter unten). Mit den Tasten am Glove kann das Stück gestartet, beendet oder unterbrochen werden, oder es können neue Samples geladen werden.

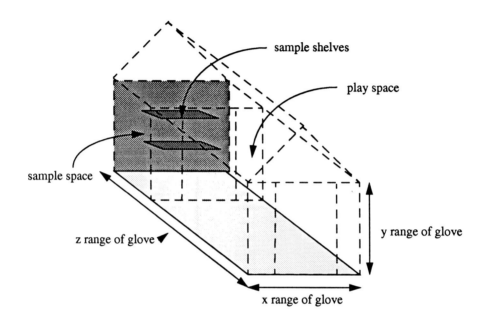

Originally intended for performance, Gate was first presented as an installation at Mills College in Oakland, California, as part of the Electronic Music Plus Festival, and subtitled UFOs. Gate (UFOs) was conceived of as an enclosed residential space spanning the fuzzy border between two worlds. A fence was constructed along the trigger plane, enclosing the sample space, with an opening for the gate. On the sample space side of the fence sat a small table. On the tabletop were about a half-dozen invisible, unidentified household objects, pots, pans, pepper grinders, salt shakers, etc., each represented by an audio sample, each at a fixed location relative to the trigger plane. While the gate was open, an object could be 'picked up' from the table by moving the glove to the object's location and putting the thumb and index finger together. The object was then brought back through the gate into play space, where its sonic properties were explored and

Gate wurde zwar ursprünglich als Live-Performance konzipiert, gelangte dann aber das erstemal als Installation im Mills College in Oakland, Kalifornien, im Rahmen des Electronic-Music-Plus-Festival an die Öffentlichkeit, und zwar mit dem Untertitel „UFOs". Gate (UFOs) war als geschlossener Wohnraum geplant, der die verschwimmende Grenzlinie zwischen zwei Welten überspannen sollte. Entlang der Auslöser-Ebene wurde ein Zaun errichtet, der bis auf eine Öffnung für das Gate den Sample-Raum umschloß. Auf der Sample-Raum-Seite des Zaunes stand ein kleiner Tisch. Auf dem Tisch befanden sich etwa ein halbes Dutzend unsichtbare, nicht benannte Haushaltsgegenstände – Töpfe, Pfannen, Pfeffermühlen, Salzstreuer etc. – die jeweils durch ein Audiosample dargestellt waren und sich in einer im Verhältnis zur Auslöser-Ebene feststehenden Position befanden. Wenn das Gate offen war, konnte ein Gegenstand vom Tisch „genommen" werden, indem man den Glove an die Position des Gegenstandes bewegte und Daumen und Zeigefinger aneinanderlegte. Dann brachte man den Gegenstand durch das Gate in den Spiel-Raum, wo sich seine Klangeigenschaften untersuchen und manipulieren ließen, indem man den Power Glove in seinem 3-D-Feld umherbewegte.

Haunted ... Wirklich virtuelle Immobilie

Im späten 19. und frühen 20. Jahrhundert verwendeten Magier und Medien besonders konstruierte Möbelstücke, sogenannte „Séancekabinette", als Tore zur Welt der Geister. Durch diese Kabinette kommunizierten Leute wie die Davenport Brothers oder Medien wie Mina Stinson Crandon mit den Verstorbenen. Mit Hilfe der damaligen technischen Spezialeffekte (Stimmtrompeten, pneumatisch gesteuerte Hände) und ein paar Tricks konnten diese „Spiritisten" Erscheinungen und Geister hervorzaubern. Priester und Schamanen hatten spezielle Hilfsmittel und Techniken, um in den Häusern der Lebenden wilde, lärmende Geister heraufzubeschwören und dann wieder zu vertreiben.

Wie Gate (UFOs), ein Séancekabinett oder ein Geisterhaus steht Gate (Haunted) für ein Tor zu einer anderen Welt, einer vorübergehenden Heimstatt für telenomadische Seelen, die auf die „andere Seite" hinübergewechselt sind. Statt – wie bei UFOs – auf dem Tisch, befinden sich die Samples auf Wandregalen. Durch Seile werden ein Haus und ein bestimmter Raum in dem Haus skizziert. An der Rückwand des Raumes befinden sich ein Relief mit schamanischen Symbolen und Fetischen und zwei Fächer. Auf diesen Fächern befinden sich vier rituelle Gegenstände zur Geisterbeschwörung. Die Benützerin betritt das Haus und geht auf den Raum zu. Von einem kleinen Tisch neben der Tür zu dem Raum nimmt sie einen Dataglove und steckt ihre rechte Hand hinein. Sobald sie durch die Tür zur Rückwand des Raumes greift, hört sie, wie die Tür sich öffnet. Mit der behandschuhten Hand nimmt sie einen der rituellen Gegenstände – der Gegenstand gibt einen Ton von sich. Durch die Tür holt sie den Gegenstand in den vorderen Teil des Hauses. Hier ändert sich der Klang des Gegenstandes, wenn sie ihn hin und her bewegt. Dann greift sie wieder durch die Tür, legt den Gegenstand an seinen Platz auf dem Regal zurück und nimmt einen anderen.

Seven

Als Kind hörte ich viel Kurzwellen- und Amateurradio – Radio war das Hobby meines Vaters, seit er ein Teenager war, und eine Zeitlang arbeitete er auf Hawaii auch bei einer Radiostation. Der Cheftechniker dieser Station machte das nur nebenbei und war eigentlich bei einer Firma angestellt, die für das Defense Department arbeitete. Einmal nahm er mich in der Nacht zu einer Radar-Teststation am südöstlichsten Zipfel der Insel mit, einem ganz entlegenen Ort, wo angeblich KGB-Agenten aus ihren U-Booten an Land kamen und sich

manipulated by moving the Power Glove within its 3D field.

Haunted... Real Virtual Estate

In the late 19th and early 20th century magicians and mediums used specially constructed furniture, "seance cabinets", as gateways to the spirit world. With these structures showmen such as the Davenport Brothers or mediums like Mina Stinson Crandon would communicate with the dead. Aided by the high-tech special effects of the day (voice trumpets, pneumatically controlled hands) and a little legerdemain, these "Spiritualists" could create apparitions and manifest spirits. Priests and shamans use special tools and techniques to conjure and then exorcise noisy, rambunctious, spirits from the homes of the living.

Like Gate (UFOs), a seance cabinet or a haunted house, Gate (Haunted) suggests a doorway to another world, a temporary home for telenomadic souls that have gone over to the "other side". Rather than the table used in UFOs, the samples sit on shelves mounted on a wall. Ropes outline a house and a special room within the house. The back wall of the room is decorated with a relief of shamanic symbols and fetishes and supports two shelves. On these shelves are four ritual objects used for summoning ghosts. The user enters the house, walking towards the room. She picks up a dataglove from a small table near the door to the room, and puts her right hand in the glove. Reaching through the door to the back wall of the room she hears the door open. She picks up one of the ritual objects with her gloved hand... it makes a sound. She brings the object back through the door into the rest of the house. In this part of the house the sound the object makes changes as she moves the object around. She reaches through the door again to replace the object in its proper place on the shelf, and chooses another object.

Seven

I heard a lot of shortwave and amateur radio while I was growing up... my father had been a radio hobbyist since his teenage years and was in the broadcasting business on the Big Island of Hawaii for awhile. The chief engineer who worked at my father's radio station did so as a sideline, spending most of his time working for a defense department contractor. I remember one night he took me with him to a radar test station on the southeastern tip of the island, a remote spot where, local rumor had it, KGB agents came ashore from submarines and blended in with the tourists. The station was full of electronic gear, most of it constantly on and making noise... distant AM radio stations from the mainland, shortwaves, weird military telemetry. Those sounds still resonate for me today.

Seven Gates is an interactive computer music composition, the live performance variant of the Gate series of installations. The performer uses the glove to manipulate invisible "sonic souvenirs", audio samples from the various musical cultures around the Pacific Rim (imagine driving around California with your car radio stuck on "scan" mode). The piece has seven sections ranging in length from two to five minutes... either a set of "bagatelles" or a sort of "theme and variations", depending on how rigorously you want to define the latter. As in the Gate series, the stage is divided into two areas: a 'sample space' with invisible 'shelves' and a 'play space'. Separating the two areas is an invisible 'fence'. This fence runs perpendicular to the z axis at it's midpoint, for the full extents of the x and y axes. The performer, wearing the glove, reaches through a 'gate' in the fence to grab a sample from a shelf, then brings it back through the gate to the play space where its sound can be modulated. The gate is wired for sound; anytime it's opened or closed an audio sample (one of several gates and doors) is played. (Each of the seven sections has a different sample of a gate or door associated with it.) While the performer holds the sample (by making a fist) the sample is audible; making the hand flat mutes the sample.

The piece is mostly improvisational, with a few general rules about gestures to use in each section. These rules take the form of seven simple diagrams that define how my gloved hand moves within sample space and play space (see below). The repertoire of gestures used in the piece is symmetrical: sections one and seven use the same or similar gestures, sections two and six the same, and so on. The central fourth section generally matches the first and seventh... I've been known to deviate from my 'score' during this part. There's a contextual and timbral symmetry at work in Seven Gates as well. The first and seventh sections use samples of Nortena and Pacific Island music, the second and sixth use 18th century European music, the third and fifth use speech sampled from radio and television, and the fourth section is comprised of a combination of the above.

unter die Touristen mischten. Die Station war voll elektronischer Geräte, von denen die meisten immer eingeschaltet blieben und irgendwelche Geräusche machten – weit entfernte Mittelwelle-Radiostationen auf dem Festland, Kurzwellen, unheimliche militärische Meßgeräte. Diese Geräusche klingen mir heute noch in den Ohren.

Seven Gates ist eine interaktive Computermusik-Komposition, die Bühnen-Live-Version der Gate-Installationen. Mit dem Glove manipuliert der Künstler unsichtbare „Klangsouvenirs", Audiosamples verschiedener musikalischer Kulturen am Pazifik (wie wenn man durch Kalifornien fährt und sein Autoradio auf „Scan" eingestellt hat). Das Stück hat sieben, zwischen zwei und fünf Minuten lange, Teile – eine Folge von „Bagatellen" oder eine Art „Thema und Variationen", je nachdem, wie streng man letzteres definieren will. Wie auch bei den Gate-Installationen ist die Bühne in zwei Bereiche geteilt: einen „Sample-Raum" mit unsichtbaren „Regalen" und einen „Spiel-Raum". Die beiden Räume werden durch einen unsichtbaren „Zaun" getrennt. Dieser Zaun verläuft senkrecht zur z-Achse an deren Mittelpunkt und erstreckt sich über die gesamte Länge der x- und y-Achsen. Der Künstler greift mit dem Handschuh durch ein Gate, ein Tor, im Zaun, holt ein Sample vom Regal und bringt es durch das Tor in den Spiel-Raum, wo sein Klang dann moduliert werden kann. Das Tor ist klang-verdrahtet, und jedesmal, wenn es geöffnet oder geschlossen wird, wird ein Audiosample gespielt. (Mit jedem der sieben Teile ist ein anderes Gate- oder Torsample assoziiert.) Solange der Künstler das Sample festhält (indem er eine Faust macht), ist es hörbar, macht er die Hand auf, verstummt das Sample.

Das Stück basiert zum Großteil auf Improvisation, abgesehen von ein paar allgemeinen Regeln hinsichtlich der in jedem Teil etwas unterschiedlichen Handbewegungen. Diese Regeln sind sieben einfache Diagramme, die die Bewegungen der behandschuhten Hand im Sample-Raum und im Spiel-Raum definieren (siehe unten). Das Repertoire der in diesem Stück verwendeten Handbewegungen ist symmetrisch: die Teile eins und sieben haben dieselben oder ähnliche Handbewegungen, ebenso die Teile zwei und sechs, und so fort. Der zentrale, vierte Teil entspricht normalerweise den Teilen eins und sieben – ich weiche bei diesem Teil allerdings meist von meiner „Partitur" ab. Seven Gates weist außerdem eine kontextuelle Symmetrie und eine Klangfarbensymmetrie auf. Die Teile eins und sieben verwenden Samples aus der Musik der pazifischen Inseln und der Nortena-Musik, die Teile zwei und sechs europäische Musik aus dem 18. Jahrhundert, die Teile drei und fünf Sprechsamples aus Radio und Fernsehen, und im vierten Teil werden die oben genannten Samples miteinander kombiniert.

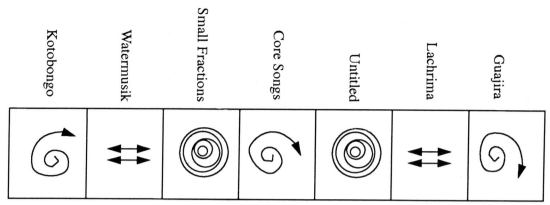

Manchmal werde ich von Leuten, die Seven Gates auf der Bühne gesehen haben, gefragt, ob ich Tai-Chi mache oder irgendeine Tanztherapie. Die Antwort ist ein bedingtes Nein, obwohl es nach den vielen Jahren, in denen ich nur Tasten gedrückt und Computermäuse umhergeschoben habe, gut tut, auf der Bühne einmal richtig ins Schwitzen zu kommen. Die Beziehung zwischen Klang und Bewegung ist einfach und direkt, der Klang gibt den Ton an. Die Handbewegungen entstehen dadurch, daß große Brocken auditiver kultureller Verweise genommen und zu rein elektronischen Klangfarben zusammengeschrumpft werden – das alltägliche Klangterrain wird auseinandergenommen, wieder zusammengebaut und neu angelegt.

Schluß

Einige Definitionen des Wortes „gate" aus dem American Heritage Dictionary: „etwas, das Zutritt gewährt"; „eine Öffnung in einer Wand oder einem Zaun als Eingang oder Ausgang"; „die Konstruktion, die so eine Öffnung umschließt". In jedem Stück der Gate-Serie werden ein paar gefundene Sounds genau untersucht. Mit meiner selbstgebrauten Software in Verbindung mit dem Glove kann der Künstler in das Innere dieser im elektronischen Gedächtnis gefriergetrockneten Klänge vordringen. Der Zuschauer oder Installationsbenutzer taucht nicht in eine auditiv oder visuell realistische Welt ein; durch die Verwendung gesampelter Sounds und ein bißchen Bühnenbilddesign wird seine Phantasie miteinbezogen. In den Installationen Gate (UFOs) und Gate (Haunted) definieren Darstellungen von Zäunen und Wänden einen Zugang zu dieser imaginären Welt. Indem der Benutzer die Einladung einzutreten, annimmt, wird er zum Künstler und agiert als Medium – er erweckt Klangerscheinungen zum Leben, tritt in ungehörte Welten ein.

Sometimes people who've seen a performance of Seven Gates ask me if I've been studying Tai Chi or practicing some kind of alternative dance therapy. The answer is a qualified no, although after years of tweaking knobs and moving computer mice it's refreshing to actually perspire during a performance. The relationship between sound and movement is simple and straightforward, and the sound is in charge. The gestures in the piece are the results of grabbing big chunks of audible cultural references and fluidly shrinking them down to purely electronic timbres... disassembling, reconstructing, and remapping everyday sonic terrain.

Conclusion

Some definitions of the word 'gate' from the American Heritage Dictionary: "Something that gives access","An opening in a wall or fence for entrance or exit", "The structure surrounding such an opening". Each of the pieces in the Gate series is a detailed exploration of a small set of found sounds. My home brewed software teamed with the glove gives the performer access to the insides of those sounds, freeze-dried in electronic memory. Instead of immersing the audience member or installation user in an aurally and visually realistic world, I use sampled sounds and a bit of theatrical set design to bring his or her imagination into play. In the installations UFOs and Haunted representations of fences and walls identify an entrance to this imaginary world. Accepting the invitation to step inside, the performer becomes a performer and acts as medium... reanimating sonic apparitions, accessing unheard worlds.

Contact: Mark Trayle
POB 192014, San Francisco, CA
94119-2014 USA
tel 01.510.527.9136, fax 01.510.527.7247
email met@well.sf.ca.us

FLOW

CHARLY STEIGER

Flow Data

Media systems: the collection and distribution of data, discoveries and the administration of runtimes and speeds, the measurement and bridging of distances. We believe that we are going in one direction, our goal in sight – but which way is the river of data flowing? Which systems are connected? Where is their input; where is their output? Who is in control? Who is interpreting?

An important area of interest in media technologies has focused on the linkage of various systems: How can television and interactive videotext be coupled, what about language recognition and telecommunications, assembly line inspection and the improved adjustment of robot efficiency? Connections, links – interruption, disconnection – the question of what the axioms are, which rules of linkage they define, and how they can be imagined, intended and conceptualized becomes more and more urgent. Do data-processing technologies lead solely to the creation and duplication of artificial worlds of images, or can a conceptualization arrived at from images of social relevance precede the systems? And: who will achieve this conceptualization? Which images will precede it?

At present, the world now being created is even less able to produce its interpretation on the basis of a singular object than before. Now, the object can no longer be considered the only example of its type (media or genetic) with certainty, our behavior, our perception at the location of the putative or actual event (situ) becomes again the object of active interest.

For this reason, the artistic problem of representing the world concentrates only in exceptions on the problem of using an object to represent a concept. On the contrary, the creation of situations which demand the description of pointedly chosen systems within systems superordinate to them by means of adjustment and contrast is the more important issue.

Flow – Daten

Mediale Systeme: Sammeln und Ausgeben von Daten, Feststellen und Verwalten von Laufzeiten und Geschwindigkeiten, Messen und Überbrücken von Distanzen. Unser Ziel vor Augen glauben wir in eine Richtung zu gehen – doch wie fließt der Datenstrom? Welche Systeme sind gekoppelt? Wo ist ihre Eingabe, wo die Ausgabe? Wer kontrolliert? Wer interpretiert?

Ein wesentliches Interessensgebiet medialer Technik hat sich der Verknüpfung verschiedener Systeme zugewandt: Wie kann Fernsehen mit BTX, wie Spracherkennung mit Telekommunikation, wie Bandkontrolle mit verbesserter Leistungsanpassung der Robotik verbunden werden? Verbindung, Verknüpfung – Unterbrechung, Trennung – immer dringlicher wird die Frage danach, welche Axiome welche Verknüpfungsregeln definieren, wie sie vorgestellt, gedacht und konzeptualisiert werden können. Führen die Datenverarbeitungstechniken lediglich dazu, daß artifizielle Bildwelten entstehen und vervielfältigt werden, oder kann den Systemen eine an den Bildern sozialer Relevanz gewonnene Konzeptualisierung vorausgehen? Und: wer leistet diese Konzeptualisierung? Welche Bilder gehen ihr voraus?

Die Welt, die im Entstehen ist, kann weniger denn je am singulären Objekt ihre Interpretation leisten – dort wo das Objekt nicht mehr zweifellos als das einzige seiner Gattung (medial oder genetisch) gelten kann, wird unser Verhalten, unsere Wahrnehmung am Ort des vermeintlichen oder tatsächlichen Geschehens (situ) aufs neue Gegenstand wirksamen Interesses.

Das künstlerische Problem der Darstellung von Welt konzentriert sich deshalb nur noch in Ausnahmefällen auf die Problematik der Repräsentation eines Konzepts in einem Objekt. Vielmehr geht es darum, durch Einpassung und Kontrastierung pointiert ausgewählter Systeme in ihnen übergeordneten Systemen Situationen zu erzeugen, die ihre Beschreibung herausfordern.

Flow – Videoinstallation

Zwei Informationsbereiche: ein Projektionsbereich mit einem sechsteiligen Projektionsfries und ein darunter auf-

steigender Treppenaufgang im Brucknerhaus, der von ebenfalls sechs Lichtschranken erfaßt wird.

Jeweils ein Bewegungsmelder ist mit einem Projektionssystem, einem Videobeamer, gekoppelt, so daß, durchschreitet man den Treppenaufgang, jeweils auf der Höhe des entsprechenden Beamers die formale Präsentation der Projektion – eines zu einem Band gestauchten, mutierten Fernsehbildes – verändert wird: sie erhält einen veränderten Input und klappt – für den Moment des höhenversetzten Entlangschreitens – von der Waagerechten in die Senkrechte.

Sowohl die beobachtenden Passanten des Projektionsfrieses auf der Galerie oberhalb der Treppe als auch die Treppensteigenden werden im Moment des Ereignisses ihren Einfluß auf das System noch ungenau interpretieren. Dem Benutzer auf der Galerie wird die Kopplung von Bewegung an die Veränderung der projizierten Bilder erst nach einer gewissen Beobachtungszeit als systematische Einheit erscheinen – genauso wie der Treppensteigende erst nachdem er die Position des Beobachtenden eingenommen hat, seinen Einfluß auf die Projektionssystematik entsprechend interpretieren wird.

Die Installation legt den Begriff „Interaktion" zwischen Mensch und Gerät nicht als sofortige, unmittelbare Reiz-Reaktions-Systematik aus, sondern als eine mehrschichtige Verkopplung von technischer Kontrolle und menschlicher Beobachtung, von unmittelbarer Bearbeitung elektronischer Daten und den auf Zeit und Interpretation basierenden Prozessen der Selbst- und Außenwahrnehmung. Denn erst die wechselseitige Wahrnehmung jeweils anderer Informationsebenen führt zum Verständnis des Gesamtsystems und der eigenen Funktion in ihm.

CHARLY STEIGER / ACHIM WOLLSCHEID

Ich danke Mike Krebs von TVT, Frankfurt, Bernd Geisler von viteg, Neu-Anspach und Delta-System Studioanlagen, Rödermark für die freundliche Unterstützung dieser Arbeit.

Flow Video Installation

Two areas of information: a projection area with a six-part projected frieze and, below it, a stairway located in the Brucknerhaus which is equipped with six light barriers.

Each movement sensor is connected to a separate video projector so that when one steps onto the stairway, the formal presentation of the projection – a mutated television image which has been compressed into a band – is changed according to the height of the corresponding projector: This projection contains an altered input and swings – for the duration of the step – from a horizontal to a vertical position.

Both the people passing by the projected frieze on the gallery above the staircase and the person on the stairs will be unable to clearly interpret their effect on the system at the moment that the event occurs. The coupling of movement to the change in the projected images will be obvious to the user on the gallery as a systematic unit after a certain time of observation – just as the person on the stairs will interpret his or her influence on the projection system after having assumed the position of observer.

The installation does not interpret the concept of "interaction" between man and machine as an instantaneous, direct system of stimulus and reaction, but as a multi-layered coupling of technical control and human observation, of the immediate processing of electronic data and processes of observing the self and the environment, which is based on time and interpretation. Understanding the entire system and an individual's own function within it is not possible without mutual perception of differing levels of information.

I would like to thank Mike Krebs from TVT, Frankfurt; Bernd Geisler from viteg, Neu-Anspach and Delta-System Studioanlagen, Rödermark for their kind help.

Cinema Spike

KARIN HAZELWANDER

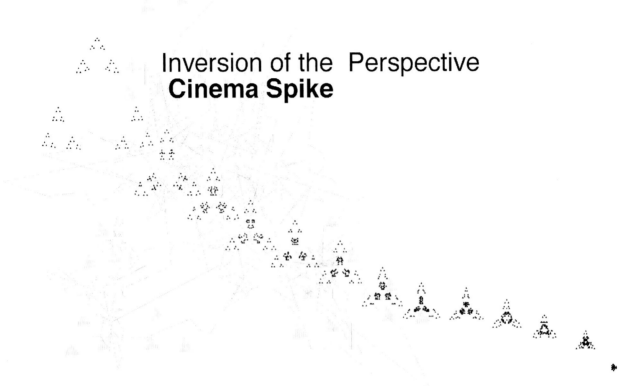

Inversion of the Perspective
Cinema Spike

Perception:

"Cinema Spike" is a steel sculpture which communicates with the viewer and interacts with space and with sight. It communicates within itself (see Sketch 3), with itself and also with the outside world, as soon as a viewer approaches it. Without these viewers – human beings – it cannot produce worlds of images, the perceptible result of which could be documented. In communicating with humans, the area in which the communication takes place is always one-sided. The sensory or emotional re-action is reserved for the human.

Technique:

The space is formed by three-sided columns, and a digital image is projected onto each of the three sides. The special arrangement of a fractal triangle enables interaction between the viewer

Wahrnehmung:

Cinema Spike ist eine mit dem Betrachter kommunizierende, mit dem Raum und dem Sehen interagierende Stahlskulptur. Sie kommuniziert in sich selbst (siehe Skizze 3), mit sich selbst und darüber hinaus nach außen, sobald ein Betrachter auf sie zutritt.
Ohne diesen Betrachter – den Menschen – kann sie keine Bildwelten erzeugen, deren wahrnehmbares Resultat dokumentierbar wäre. In der Kommunikation mit dem Menschen bleibt der Kommunikationsraum einseitig gerichtet. Die sinnliche oder emotionale Re-Aktion bleibt dem Menschen vorbehalten.

Technik:

Der Raum wird gebildet aus dreikantigen Säulenkörpern, auf deren jede Seite ein digitales Bild projiziert wird. Die spezielle Anordnung eines fraktalen Dreiecks ermöglicht eine Wechselwirkung zwischen dem Betrachter und zwei der drei Bildseiten. Insgesamt wird die Projektion an 81 Stahlsäulen vorgenommen. Die Projektion wird maltechnisch umgesetzt. Um ein identisches Bild zu sehen, muß der Beobachter in ca. 5 Meter Entfernung stehen. Die gesamte Länge der Skulptur im Grundriß beträgt 2,43 m. Die Summe von Säulen und Bildern ergibt sich aus der Anzahl der Pixel eines Monitorbildes und der fraktalen Berechnung eines Images in diesem dreidimensionalen Kommunikationsraum.

Reflexion:

Regulative Prinzipien abstrakter Gedankengänge werden anhand einer bildlichen Umsetzung durch diese Arbeit verdeutlicht. Anhand von 2 Images, dem halb abgewandten Gesicht einer Frau und dem eines Mannes, wird die sterile Kommunikation als soziales Spiel dargestellt, sprachlich kommentiert durch das Wort: „Dialog" Auf daß eine „Intelligent Ambience", die auch als schlichter Luxus interpretiert werden kann, auch zur „Ambient Intelligence" führt, die für die menschliche Entwicklung von essentieller Bedeutung ist, muß die „Wechselwirkung" betont werden. Dies ist das zentrale Thema dieser Arbeit.

Wechselwirkung:

Wo liegt der Schwerpunkt einer Wechselwirkung?
Kann Vertrauen Objektivität ersetzen?
Welche Farbe hat der Maßstab?
All diese Fragen werden beim Eindringen in die Kommunikationsebene „Cinema Spike" beantwortet.

Cinema Spike von Karin Hazelwander und Wolfgang Werner in wissenschaftlicher Zusammenarbeit.
Besonderen Dank an Dr. Michael Schmid (Institut für Allgemeine Physik, TU Wien), an Ebenberger Wien (der graue Rabe fliegt) und an AMAG (Aluminium Ranshofen Presswerk).

and two of the three sides. In all, images will be projected onto 81 steel columns. The projection will make use of painting techniques. In order to see an identical image, the viewer must stand at a distance of approximately five meters. The entire length of the sculpture' layout is 2.43 m.
The sum total of the columns and images is the result of the number of pixels in a monitor image and the fractal calculation of an image in this three-dimensional communication space.

Reflection:

This work illustrates regulative principles of abstract trains of thought on the basis of a pictorial realization. On the basis of two images, the faces of a woman and a man which are half turned away, sterile communication is represented as a social game, commented upon linguistically with the word "dialog". So that an "intelligent ambience", which can also be interpreted as plain luxury, is able to lead to an "ambient intelligence" of essential importance to human development, "interaction" must be stressed. This is the work's central theme.

Interaction:

Where can the focus of an interaction be found?
Can trust replace objectivity?
Which color is the standard?
All of these questions will be answered after entering the "Cinema Spike" communication space.

"Cinema Spike": by Karin Hazelwander and Wolfgang Werner in scientific cooperation.
Special thanks to Dr. Michael Schmid (Institute of General Physics, TU Vienna), to Ebenberger (the grey raven flies) and to AMAG (Aluminum Ranshofen Squeezer).

sabotage™ XIII

ROBERT JELINEK

(Foto: Christa Radl)

SABOTAGE XIIIa / 1993, Management Club, Wien

In the "Management Club" in Vienna a number of microphones were installed in the room. The discussions and lectures of those present were transmitted into the public.

SABOTAGE XIIIa / 1993, Management Club, Wien

Im Wiener „Management Club" wurden zahlreiche Standmikrophone im Raum plaziert. Die Gespräche und Vorträge der Anwesenden wurden in den öffentlichen Raum übertragen.

(Foto: Standa Vana)

**SABOTAGE XIIIb / 1993,
Staromestnì N·mestì, Prag**

Auf einem öffentlichen Platz wurde ein Mikrophon ohne Strom- und Lautsprecheranschluß aufgestellt.

**SABOTAGE XIIIb / 1993,
Staromestnì N·mestì, Prag**

A microphone with no electricity or loudspeaker connection was installed in a public square.

(Foto: Norbert Artner)

SABOTAGE XIIIc / 1994, Ars-Electronica, Linz

In the rooms of the Brucknerhaus in Linz flowers in flowerpots were installed in varying distances. Microphones (bugs) and loudspeakers were hidden in the artifical flowers. The conversations between the visitors standing close to the pots were recorded and transmitted at the same time.

SABOTAGE XIIIc / 1994, Ars-Electronica, Linz

In den Räumlichkeiten des Linzer Brucknerhauses wurden in verschiedenen Abständen vier Blumentöpfe samt Blumen am Boden installiert. In die Blüten der Kunststoffblumen waren Mikrophone (Abhörwanzen) und Lautsprecher eingebaut. Die Gespräche der Besucher, die sich in unmittelbarer Nähe der Blumentöpfe befanden, wurden gleichzeitig abgehört und übertragen.

ATLANTIS CONSTRUCT

FÜRST – THALER

die muse der bildenden künstler heißt nicht muse, sie heißt techne.
t.w.a.

happy nautis

– und dort wogten gewaltige sternensegler glimmend, zittrig – trostlos auf der suche nach helden, nach taten, nach reichtümern, nach menschen.
hastig zwischen exotisch klingenden sonnen in den entlegenen provinzen der peripheren galaktischen langeweile.
in diesen tagen war der mut wille, noch ungebrochen – das risiko noch hoch und richtige männer, richtige frauen noch und kleine happy nautis noch richtige kleine happy nautis.
entweder war das klima in dieser spätnachmittagsdekade nicht ganz so, wie es sein sollte – der tag war lang, eine halbe stunde lang – oder in den unüberschaubaren wellen der kristalltierchen lagen bilder verbotener biogener wünsche.
– der mereskristall hatte und bekam nicht das richtige beruhigend weiche rosa.
– blau.
so entstand im irgendwo des galaktischen museums – terra – die humanoide kondition für einen phantastisch primitiven, neuen industriezweig. die genetische verwirrung happy nautis. haustiere mit eigenen und fremden programmen, weich und hart, lumineszent, trivial, neugierig und spiegelverrückt.
blau,
blau nach sonderwünschen,
blauer strom
und kein rosa mehr ...

the name of the artist's muse is not muse; she is called techne.
t.w.a.

happy nautis

- and powerful starships surged there, gleaming, trembling – disconsolately searching for heroes, deeds, for riches, for humans.
hurrying between exotic sounding suns in the distant provinces of the peripheral galactic boredom. courage was resolute, still unbroken in those days – the risk was still great and real men, real women still existed, and happy little nautis were still real, happy little nautis.
either the climate was not quite that which it should have been on that late afternoon decade – the day was long, half an hour long – or images of forbidden biogenous wishes lay in the unimaginable waves of the little crystalline animals.
- the marine crystal did not have and was not given the correct, soothing soft pink.
- blue.
this is how, somewhere in the galactic museum – terra – the humanoid condition was created for a fantastically primitive, new industry. the genetic confusion of happy nautis. pets with their own and foreign programs, soft and hard, luminescent, trivial, curious and mirror-crazy.
blue,
blue by special requests,
a blue current
and no more pink ...

happy nautis monitor objekte 1994

material	schaumstoff, silicon, edelstahl
maß	2,3 meter
team	christoph fürst v. freystadt
	harwald v. hatschenberg
	werner kramer
	michael pointner
	gerold andreas g. thaler-hohenfels
cover	johannes domsich

besonderen dank für die freundliche unterstützung des projektes: dr. peter greiner, fa. greiner schaumstoffe österreich; fa. dow cornig silikon austria.

happy nautis monitor objects, 1994

material	foam, silicone, high-grade steel
size	2.3 meters
team	christoph fürst v. freystadt
	harwald v. hatschenberg
	werner kramer
	michael pointner
	gerold andreas g. thaler-hohenfels
cover	johannes domsich

ROBIN HOOD INC., wissenschaftlich forschende kunst

TRIGGER YOUR TEXT

ELFRIEDE JELINEK
GOTTFRIED HÜNGSBERG
HANNES FRANZ

Two 3-meter high sweeping silhouettes stand in the room as counterparts to one another. While the rear form is leaning against the wall, the front one is free-standing. Both squat aluminum figurines are covered with a sensitive orange-red polyamide sheath which was applied with electrostatic flocking. The silhouette-like structure awakens a wide variety of associations of an anthropomorphic nature, from Baroque altar fronts to Far Eastern shrines. The shape of the meter-high cylinder placed in front of them, the front of which is equipped with an operating lever, reminds one of a candle. This cylinder contributes a great deal to the hieratic and religious character of the installation.

Using a joystick connected to one of the monitors built into the silhouette forms, the visitor can play an aggressive video game based on the destruction of virtual flies. Depending on how successful one is at swatting flies, the player can call a selection from three chapters in Elfriede Jelinek's book *Wolken.Heim.* on the second monitor. By increasing the level of difficulty, one can advance to other levels of the text, which are

organized according to the system of German Idealism – the "pure" idea, the "political" idea and the "practical" idea.

"The experiences which I have had in the theater in the past few years have made me think about the basics of distributing literature. Because of my involvement with computer art,

Zwei 3 m hohe, geschwungene Silhouettenformen stehen als Pendants zueinander versetzt im Raum. Während die hintere Front an der Wand lehnt, ist die vordere freistehend plaziert. Die beiden gedrungenen Aluminiumfigurinen sind mit einer empfindlichen, orange-roten Polyamidbeschichtung überzogen, die durch elektrostatische Beflockung aufgetragen wurde. Die scherenschnittartigen Gebilde wecken die unterschiedlichsten Assoziationen, von antropomorphen Anklängen über barocke Altarprospekte bis hin zu fernöstlichen Schreinen. Der davor aufgestellte, meterhohe Zylinder, auf dem ein Bedienungshebel angebracht ist, erinnert formal an eine Kerze. Dies trägt in entscheidendem Maße zu dem hieratischen und sakralen Charakter der Installation bei.

Über den Joystick kann der Betrachter ein aggressives Videospiel, welches auf das Totschlagen virtueller Fliegen abzielt, auf einem der beiden in die Silhouettenform eingebauten Monitore steuern. Je nachdem, wie mit der Fliegenklatsche umgegangen wird, gelingt es dem Spieler, eine Auswahl von drei Kapiteln aus Elfriede Jelineks Buch *Wolken.Heim.* auf dem zweiten Monitor abzurufen. Die Steigerung des Schwierigkeitsgrades ermöglicht es, in weitere Textebenen vorzudringen, die nach dem System des deutschen Idealismus aufgebaut sind – der „reinen" Idee, der „politischen" Idee und der „praktischen " Idee.

„Meine Theatererfahrungen der letzten Jahre haben mich dazu gebracht, grundsätzlich über die Vermittlung von Literatur nachzudenken. Durch meine Beschäftigung mit Computerkunst, überhaupt mit den neuen elektronischen Medien, bin ich zu dem Schluß gekommen, daß neue Formen der Literatur gefunden werden müssen, in denen der Rezipient von Kunst selbst auch in das Material, das ihm vermittelt wird, eingreifen können muß. Das Kunstwerk soll seine eigene Rezeption bereits enthalten! Und gleichzeitig wird der Betrachter ein Teil des Werks."
(E. Jelinek)

Trigger Your Text

Die Installation „Trigger Your Text" enthält ein Computerspiel, in dem der Spieler Texte aus Wolken.Heim.* gewinnen kann. Der Rechner hat 128 Textfragmente aus Wolken.Heim. gespeichert, die in drei Gruppen geordnet sind:
1. Die „reine Idee"
2. Die „politische Idee"
3. Die „praktische Idee"

Die drei Gruppen werden jeweils in drei 'Härtegrade' unterteilt, so daß neun Gruppen zur Auswahl stehen.

Die Installation stellt sich im Vorspann vor, fordert dann zum Spielen auf und zeigt eine kleine Gebrauchsanleitung an.

Das Spiel besteht in 'Fliegenklatschen'. Man muß versuchen, mit dem Joystick möglichst viele Fliegen zu erschlagen, ohne von den Fliegen in die Ecke gedrängt zu werden.

Der Gewinn besteht aus *Wolken.Heim.*-Texten, die von verschiedenen Schauspielern gesprochen oder gesungen werden. Welchen Text man gewinnt, hängt davon ab, wie man gespielt hat.

Falls keiner spielt, zeigt die Installation Videoclips und Textbeispiele und lädt ab und zu zum Spielen ein.

© 1993: Elfriede Jelinek/Hannes Franz/Gottfried Hüngsberg Produziert für Literatur + Medien, 1993.

* *Wolken.Heim.*, Elfriede Jelinek. Steidl Verlag 1990. Die verwendeten Texte sind unter anderem von: Hölderlin, Hegel, Heidegger, Fichte, Kleist und aus Briefen der RAF von 1973-1976.

with the new electronic media in general, I have concluded that one must find new literary forms with which the consumer of art must be able to interact with the material with which he or she is provided. The work should already include its own reception! And at the same time, the viewer becomes a part of the work." (E. Jelinek)

Trigger Your Text

The installation "Trigger Your Text" contains a computer game in which the player can win texts from *Wolken.Heim.* *

One hundred and twenty-eight excerpts from *Wolken.Heim.* are stored in the computer, and these excerpts are organized into three groups:
1. the "pure idea"
2. the "political idea"
3. the "practical idea"

The three groups will be subdivided into three "levels of difficulty" so that nine different groups are available.

The installation introduces itself in the lead-in, requests the player to start and displays brief instructions for its use.

The game's object is to "swat flies". With the aid of the joystick. the player must attempt to kill as many flies as possible without letting them back him or her into a corner.

The prize is text excerpts from Wolken.Heim, which are recited or sung by various actors. Which excerpt one wins depends on how well one has played.

When no one is playing, the installation will show video clips and examples of the text and will occasionally invite passers-by to play.

© 1993: Elfriede Jelinek/Hannes Franz/Gottfried Hüngsberg
Produced for Literatur + Medien, 1993.

* *Wolken.Heim.*, Elfriede Jelinek, Steidl Verlag, 1990. The texts used are by Hölderlin, Hegel, Heidegger, Fichte and Kleist, etc. and are taken from the letters of the Rote Armee Fraktion, 1973-76.

SECRET OF LIFE

CONSTANZE RUHM / PETER SANDBICHLER

Extensions

Tents

The tent is transparent, flexible and mobile, and it can be disassembled. It is a module which is compatible with itself. Its dimensions are variable, and its growth is independent of the conditions and circumstances of its environment. Regarded from the exterior, it could be either a container or a sculpture. When one enters it, it becomes a container, a projection room, or an archive (which preserves many different things). It reflects the nomadic behavior of a visitor and the temporary nature of his or her presence. The tent's skin is permeable to light. One can be seen as a shadow by someone outside the tent. The only source of light within the room is a projection. For this reason, the lighting changes constantly. Therefore, the structures are temporary, and the room tends to be in a state of perpetual dissolution and transformation.

Shelves

Shelves are mounted along a side wall of the tent. These shelves are empty. It is up to the visitor to fill them, to fill them with life, to make determinations. These shelves are directories, or trajectories: pathways and also hierarchies. They are the material correspondence to and prerequisite for potential archives. A shelf is an object of the everyday world, an interlocking structure consisting of an arbitrary number of compartments and subdivisions, a container for memory and forgetting, for privacy, and for externalized human memory: a sentimental object in all its functionality. At the same time, it is a thing which can be subjected to potential growth, which invents its own form and its own room within a room, the compartments and departments of which branch off again and again.

Extensionen

Zelte

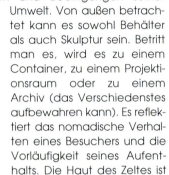

Das Zelt ist transparent, zerlegbar, beweglich und mobil. Es ist ein Modul, das mit sich selbst kompatibel ist. Seine Dimensionen sind variabel, und sein Wachstum abhängig von den Gegebenheiten und Bedingungen seiner Umwelt. Von außen betrachtet kann es sowohl Behälter als auch Skulptur sein. Betritt man es, wird es zu einem Container, zu einem Projektionsraum oder zu einem Archiv (das Verschiedenstes aufbewahren kann). Es reflektiert das nomadische Verhalten eines Besuchers und die Vorläufigkeit seines Aufenthalts. Die Haut des Zeltes ist lichtdurchlässig. Man ist für eine Person, die sich außerhalb des Zeltes befindet, als Schatten sichtbar. Die einzige Lichtquelle innerhalb des Raumes ist eine Projektion. Daher verändert sich die Beleuchtung unausgesetzt. Die Strukturen sind also vorläufig, und der Raum befindet sich tendenziell in einem Zustand dauernder Auflösung und Transformation

Regale

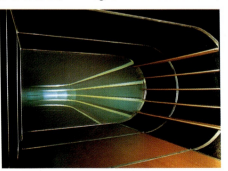

Über die Länge einer Seitenwand des Zeltes sind Regale verspannt. Diese Regale sind leer. Es bleibt dem Besucher überlassen, sie zu füllen, zu beleben, zu bestimmen. Diese Regale sind directories, oder trajectories: Pfade und Wege, und auch Hierarchien. Sie sind die materielle Entsprechung und Voraussetzung für mögliche Archive. Ein Regal ist ein Gegenstand der Alltagswelt, ein verschachteltes Gebilde, das aus einer beliebigen Anzahl von Fächern und Unterteilungen besteht, ein Behälter des Vergessens und der Erinnerung, der Privatheit, des nach außen verlegten menschlichen Gedächtnisses: in all seiner Funktionalität ein sentimentales Objekt. Gleichzeitig ist es auch ein Ding, das einem möglichen Wachstum ausgesetzt sein kann, seine eigene Form erfindet und einen eigenen Raum im Raum, dessen Ab- und Unterteilungen sich immer weiter verzweigen.

Projektionen

Das Regal, das in einem Raum leer bleibt, wird in einem anderen, künstlichen Raum benützt. Das latente Potential des wirklichen Raums, der sich durch den virtuellen Eingriff eines Betrachters (Reflexion) hypothetisch verändert, wird in der Projektion virulent. Das Licht, das von der Projektion abgegeben wird, ist nicht nur Beleuchtung im materiellen Sinn, sondern macht auch Verschiebungen wahrnehmbar. Der künstliche Raum (weil er variabler ist als der wirkliche, und seine Parameter austauschbar sind) kann leichter sichtbar werden als Projektion, die die Stabilität der realen Architektur benützt, um dem eben Festgelegten zu widersprechen. Dieser Schein-Raum ist ebenso leer wie sein wirkliches Gegenstück, trotzdem bewahrt er Dinge auf, oder er vereinheitlicht sich zu einer bewegten Skulptur, die unbegehbar, aber umso sichtbarer ist. Es ist unmöglich zu entscheiden, welcher der beiden Räume zuerst existiert hat.

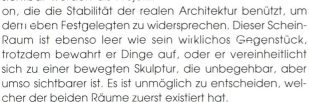

Versatiles
Kristalle

Ein Kristall besteht aus identischen Einheiten, die nach einer regelmäßigen, periodischen Struktur angeordnet sind. Kristalle symbolisieren: Perfektion, Schönheit, Symmetrie, Dauerhaftigkeit. Die Zukunft erfährt man, indem man eine Kristallkugel befragt. Ein Kristall könnte als morphologische Metapher bezeichnet werden. Früher war man der Meinung, daß Kristalle lebendig wären und wachsen könnten wie Pflanzen. Man suchte Entsprechungen zwischen der Regelmäßigkeit kristalliner Formen und den Regelmäßigkeiten bei den komplexen Wachstumsprozessen pflanzlicher und tierischer Morphogenese. Der Schlüssel zu jeder kristallinen Struktur sind Symmetrie und Symmetriebrechung. Symmetrie wird nicht als statisches System aufgefaßt, sondern als eine Anzahl von Transformationen, die zu Gruppen zusammengefaßt werden. Eine Gruppe ist ein geschlossenes Set von Transformationen.

Quasikristalle

Ein Quasikristall unterscheidet sich in seinem Aufbau auf mehrere Weisen von einem normalen Kristall. Er weist einerseits eine Art von Ordnung auf, wie sie für Kristalle typisch ist, zeigt aber zugleich eine Symmetrie, die für jede kristalline Substanz unmöglich ist. Er verkörpert eine neue Art von Ordnung, die weder kristallin noch völ-

Projections

The shelf which remains empty in one room will be used in another, artificial room. The latent potential of the real room, which is hypothetically changed by the virtual action of a visitor (reflection), becomes virulent in the projection. The light given off by the projection is not only light in a material sense; it also makes shifts perceptible. The artificial room (because it is more variable than real one, and its parameters are exchangeable) can be made visible, more easily: as a projection which utilizes the stability of the real architecture in order to refute that which has just been established. This illusory room is just as empty as its real counterpart; on the other hand, it contains things or standardizes itself into an animated sculpture which cannot be entered, but which is all the more visible. Deciding which of the two rooms existed first is impossible.

Versatile Things
Crystals

A crystal consists of identical units which are arranged according to a regular periodical structure. Crystals symbolize perfection, beauty, symmetry and durability. One can learn the future by asking a crystal. A crystal could be termed a morphological metaphor. Earlier, people believed that crystals were alive and could grow like plants. One searched for correspondences between the regularity of crystalline forms and the regularities of the complex growth processes of vegetable and animal morphogenesis. The key to every crystalline structure is symmetry and the refraction of symmetry. Symmetry is not understood to be a static system, but as a number of transformations which are combined into groups. One group is a closed set of transformations.

Quasi-Crystals

Regarding its structure, a quasi-crystal differs from a normal crystal in many ways. On the one hand, it possesses the kind of order which is typical for crystals; on the other hand, it has at the same time a symmetry which is impossible for a crystalline substance. It embodies a new kind of order which is neither crystalline nor completely amorphous. While the elementary cells of many crystals are based on platonic bodies such as cubes or octahedrons, the basis of a quasi-crystal's cells is the icosahedron. An icosahedron consists of 20 equilateral triangles and produces a so-called "five-number" * symmetry (five planes meet at every corner). For this reason, the icosahedron was not considered to be a possible elementary cell for crystalline structures until the discovery of the quasi-crystal.

Penrose-Patterns, Tilings

Penrose patterns are the two-dimensional counterparts of quasi-crystals. The pattern is not periodic; it cannot be broken down into a single elementary cell which repeats itself infinitely. However, it does fulfill certain criteria for classifying periodic mosaics composed of a basic figure (parquet patterns). In contrast to all other possible periodic mosaics, the Penrose pattern possesses a sort of five-number symmetry: In a certain sense, the pattern remains unchanged when it is turned by one-fifth of a complete revolution (72 °). Just as the Penrose pattern, the microstructure of quasi-crystals also possesses a five-number symmetry. This kind of symmetry is, however, geometrically impossible, both in periodical mosaics and in conventional materials (constructed of elementary cells which repeat regularly).

Parquet, Virus
Sletches

A new kind of order, which is found in an object belonging to the inorganic, physical world, can be described with the concept of "extension" (Deleuze). A quasi-crystal does not possess the latticed structure of crystalline forms; its structure is aperiodic, and therefore, it is a variable and, to a certain point, flexible object. In mathematics, "extension" means that a geometric object's meaning changes. An object is no longer defined as an essential form (in art, the typical essential, modern object would be sculpture, or the painted canvas), but as a variable form which spreads over a surface (comparable with viruses in a human organism, the form of which changes continuously through mutation and adaptation; or with the concept of group representation mathematics, which is based not on the form, but on transformation, i.e. movement and change; or in connection with the refraction of symmetry, the fundamental principle of which states that a sym-

symmetrische Form nach ihrer Brechung jeweils einen Zustand von mehreren möglichen einnimmt). Das moderne Konzept des Gegenstandes wird von einer prozessualen Ordnung abgelöst und erzeugt einen Gegenstand, der – ebenfalls von Deleuze – als „objectile" bezeichnet wird (object plus projectile). So könnte man die skulpturale Verwirklichung eines zuvor mittels Computer berechneten Interpolationsprozesses als „object-event" bezeichnen. Der Gegenstand beinhaltet bereits seine eigene Instabilität und die entsprechende Suche nach einem möglichen stabilen Zustand (was eine dauernde Veränderung und Ausbreitung zur Folge haben könnte). Diese Ausbreitung – ob 2- oder 3dimensionaler Art – erzeugt ein Parkett, bzw. eine Parkettierung. Eine Parkettierung ist nichts anderes als der Versuch, die Euklidische Ebene mit Vielecken zu überdecken; dabei ist eine endliche Anzahl verschieden geformter Vielecke gegeben. Diese Parkettierung ist mit Quadraten, gleichseitigen Dreiecken oder regelmäßigen Sechsecken möglich, aber nicht, wenn man regelmäßige Fünfecke verwendet. Eine Parkettierung mit Quadraten beispielsweise nennt man periodisch, da sich die Formen in exakt zwei unabhängige Richtungen wiederholen. Eine Parkettierung mit regelmäßigen Fünfecken wäre eine nicht-periodische Parkettierung. Diese nicht-periodische Parkettierung der Ebene entspricht der nicht-periodischen Struktur des Quasikristalls. So wie die Elementarzelle die kleinste Einheit eines Kristalls ist, wird der „unmögliche" Quasikristall zur kleinsten Einheit eines Systems mit labiler Balance, dessen „geordnete" Zustände nur unserem Auge geordnet erscheinen. Diese Unordnung der Dinge dringt in die Ordnung der Sprache ein: etwas scheint zu sein, was es ist, ist es aber nicht. Dieser Ansatz erscheint wieder in der extensiven Flächen- und Raumbedeckung sowohl des wirklichen als auch des virtuellen Raumes.

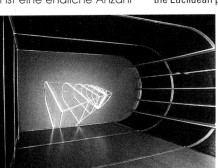

metrical form assumes one status of many possible ones after its refraction). The modern concept of objects is superseded by a procedural order and creates an object which Deleuze terms an "objectile" (object plus projectile). In this way, the sculptural realization of a process of interpolation which has previously been calculated by a computer can be termed an "object-event". The object already contains its own instability, and the corresponding search for a possible stable state (which could result in constant change and propagation). This propagation, whether two-dimensional or three-dimensional, creates a parquet, or a system of parquets. A system of parquets is nothing more than the attempt to cover the Euclidean plane with polygons. This system of parquets is possible with squares, equilateral triangles or regular hexagons, but not with regular polygons. A system of parquets with squares, for example, is termed periodic, since the forms repeat in exactly two independent directions. A system of parquets with regular polygons would be a non-periodic system of parquets. This non-periodic system of parquets on a plane corresponds to the non-periodic structure of the quasi-crystal. Just as the elementary cells are the smallest unit of a crystal, the "impossible" quasi-crystal becomes the smallest unit of a system with an unstable balance, the "ordered" state of which only seems ordered to our eye. This disorder of things penetrates the order of the language: something seems to be what it is, but it is not. This approach appears again in the extensive surface and spatial covering of both the real and the virtual space.

* „Fünfzählig" bedeutet nicht, daß die Figur, aus der der Kristall besteht, ein Fünfeck sein muß, sondern nur, daß sie nach einer Drehung um 72° nicht identisch mit sich selbst ist (so wie ein gleichschenkeliges Dreieck nach einer Drehung um 120°).

* "Five-number" does not mean that the figure comprising the crystal must be a pentagon, but that it is not identical to itself after being turned by 72 ° (as is the case with a triangle with two equal sides after being turned by 120 °).

Produktion & Support: Kay Fricke
Fotos: Kay Fricke, Christian Schoppe

Das Projekt wurde verwirklicht mit Unterstützung von:
Bundesministerium für Unterricht und Kunst, Wien
Niederösterreichische Landesregierung
Tiroler Landesregierung
Städelschule-Institut für Neue Medien, Frankfurt/Main

DER NEGIERTE RAUM

the negated room

ALBA D'URBANO

The Origin

"Pictures are meaningful surfaces. They indicate – usually – something 'out there' in space and time which they want to make conceivable for us as abstractions (as abbreviations of the four space-time dimensions on the two dimensions of the surface)" (V. Flusser: "Für eine Philosophie der Photographie", 1991).

The installation "Der negierte Raum" deals with two issues: the relationship between reality, image, text and simulation with regard to the problem of representing an architecturally defined space (which can be considered a small part of the physical spatial universe in general and spaces constructed by humans in particular). The installation's point of departure is an exhibition space "equipped" at various times with pictures or objects which have a special meaning in a cultural context. On the one hand, this is a room for presenting the installation; on the other hand, the specific spatial situation itself becomes the subject of the artistic work, the "shelter" of which it represents. It represents something and is itself represented.

Der Ursprung

„Bilder sind bedeutende Flächen. Sie deuten – zumeist – auf etwas in der Raumzeit „dort draußen", das sie uns als Abstraktionen (als Verkürzungen der vier Raumzeit-Dimensionen auf die zwei der Fläche) vorstellbar machen sollen."
(V. Flusser: „Für eine Philosophie der Photographie", 1991)

Die Installation „Der negierte Raum" thematisiert das Verhältnis von Realität, Bild, Schrift und Simulation in bezug auf das Problem der Darstellung eines architektonisch definierten Raumes, der als kleiner Teil des physikalischen Raum-Universums im allgemeinen und der von Menschen gebauten Räume im besonderen angesehen werden kann.
Ausgangspunkt der Installation ist ein Ausstellungsraum, der zu verschiedenen Zeiten mit Bildern oder Objekten „ausgestattet" wird, die im kulturellen Kontext eine besondere Wertigkeit haben. Einerseits Präsentationsort der Installation, wird die spezifische Raumsituation andererseits selbst zum Subjekt der künstlerischen Arbeit, deren „Herberge" sie darstellt: sie stellt etwas dar und wird ihrerseits dargestellt.

Die Installation läßt den Raum, indem sie einerseits kaschiert und andererseits medial visualisiert, in vielfacher Hinsicht erfahrbar werden und konstruiert somit fast ein „Portrait" der Raumarchitektur.

Die Installation wurde in verschiedenen Phasen entwickelt. Zuerst ist eine Fotodokumentation des leeren Raumes gemacht worden, während der jede Wand einzeln aufgenommen wurde. Die entstandenen Bilder sind im Rechner digitalisiert und auf verschiedene Weise verarbeitet worden. Zunächst wurden die binären Daten der Bildinhalte im Computer als Text, als ASCII-Code interpretiert, dann auf weißes Papier gedruckt und im Ausstellungsraum als Tapete appliziert. Das schwierige Verhältnis zwischen der magischen Vorstellungskraft der Bilder und der linearen Begrifflichkeit der Schrift sind im Bezug 1:1 mittels des Interfaces des Rechners direkt übesetzt und als Paradox gelöst worden.

Millionen Zeichen bedecken die Wände, den Boden und die Decke, für jede der sechs Seiten des Raumes der ihnen eigene Code-Bildinhalt. Diese Operation des Informationstransfers der vom Gegenstand abgenommenen Daten legt sich vor die Realität und verhindert den direkten Blick des Beobachters. Der Raum wird gleichzeitig negiert und hervorgehoben, in die Zeit der Vergangenheit versetzt, die nur der Erinnerung und dem Gedächtnis

The installation allows one to experience the room in which it is on the one hand concealed and on the other hand visualized by means of media in many different respects, thereby constructing almost a "portrait" of the spatial architecture.

The installation was developed in various phases. First, a photographic documentation of the empty room was made, during which every wall was photographed separately. The resulting pictures are in the computer in digital form and have been processed in various ways. At first, the binary data of the pictures' contents were interpreted in the computer as an ASCII text, printed onto white paper and hung in the exhibition space as wallpaper. The difficult relationship between the pictures' magical power of imagination and the linear, conceptual nature of the text are translated directly in a ratio of 1:1 by means of the computer's interface and solved as a paradox.

Millions of characters cover the walls, the floor and the ceiling; each of the six surfaces in the room has its own coded pictorial content. This operation of transferring information about the data obtained from the object conceals the reality and hinders a direct view. The room is simulta-

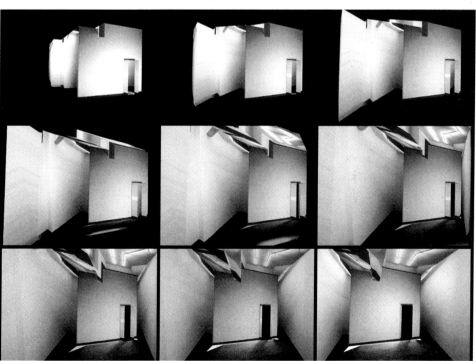

gehört: ein entmaterialisierter Raum oder eine greifbar gemachte Idee.

In diesem unverständlichen semantischen Zeichen-Ozean versunken, befindet sich im Zentrum auf einem oben offenen Podest ein Modell des Raumes aus Draht. Innerhalb des Podestes, etwas hineingeschoben, befin-

neously negated and made to stand out, placed into the past, which is a part of memory only: a room which has been robbed of its material quality and an idea which has been made concrete.

Sunk in this unintelligible semantic ocean of characters, a wire model of the room is placed in the middle on a platform which is open on top. Within the platform, shoved back somewhat, is a small rear projection screen. The viewer can look through the model and see the images of a computer animation sequence on the screen. The source of this sequence is a miniature video projector placed inside the platform and below the screen. In this video, the digitalized photographs of the room form the walls of a simulated computer model. A spatial representation is born of this combination, in which the distortion of perspective in the reproductions is mixed with the computer's exact simulation of the room, thereby producing an aspect of dislocation in the representation of the object. A synthetic room is created which ignores the normal border created by the description of Cartesian space, in which the polarizations are relativized and the definition of interior and exterior is only a question of the point of view.

The Negated Room -
- The Model World -

In the version for the "Ars Electronica '94", the actual room generated the work is done away with. Of the original object, only a few possible interpretations remain as traces: three architectural models constructed of various materials with different sizes and various qualities. They are like Russian dolls, parallel and concentric universes inside of which the computer-generated simulation is located as a fourth dimension.
The model of the language, the misunderstanding, is located on the outside – it is scaled 1:2 in relation to the original space –, and both sides of its walls are covered with ASCII characters. It resembles the miniature reproduction of the room which no longer exists and possesses the same function: That at the external shell of the entire installation. The visitor can enter this object in order to see the other parts of the work. On the inside, a closed object, a negative spatial model of the room, can be found first: the walls which surround this impermeable area consist of photographs of the interior space. A smaller wire model is added as a maximum abstraction of the architectural space's linear form. In the middle stands an open platform: The observer can look into it as if it were a fountain and let him or herself be led through the internal projection of a computer animation sequence into the non-material synthetic space of the computer.

det sich eine kleine Rückprojektions-Leinwand. Der Betrachter kann durch das Modell hineinschauen und die Bilder einer Computer-Animation sehen, die ein im Innern des Podestes aufgestellter Mini-Video-Beamer von unten auf die Leinwand projiziert. In diesem Video bilden die digitalisierten Fotos des Raumes die Flächen der Wände eines simulierten Computer-Modells. Aus dieser Zusammenstellung wird eine Raumdarstellung geboren, die die perspektivische Verzerrung der Reproduktionen mit der exakten Raumsimulation des Rechners vermischt und dadurch einen Moment der Deplazierung in der Darstellung des Objektes schafft. Es wird ein synthetischer Raum geschaffen, der die übliche Grenze der Beschreibung des kartesischen Raumes ignoriert, wo die Polarisierungen relativiert werden und die Definition von Innen und Außen nur eine Frage des Sichtpunktes ist.

Der negierte Raum -
- Die Modellwelt -

In der Version für „Ars Electronica '94" wird der reale Raum, der die Arbeit generiert hat, abgeschafft. Von dem ursprünglichen Objekt bleiben nur einige mögliche Interpretationen als Spuren: drei Architekturmodelle, die aus verschiedenen Materialen, mit verschiedenen Eigenschaften und Größen gebaut sind. Sie sind wie russische Puppen, parallele und konzentrische Universen, in deren Inneren sich als vierte Dimension die computergenerierte Simulation befindet.
Außen befindet sich das Modell der Sprache, des Mißverständnisses: Größe 1:2 im Verhältnis zum ursprünglichen Ort, seine Wände sind innen und außen von ASCII-Code-Zeichen bedeckt. Es ist wie die Miniatur-Reproduktion des nicht mehr existierenden Raumes und besitzt die gleiche Funktion: es ist die externe Hülle der ganze Installation. Der Besucher kann in dieses Objekt eindringen, um die anderen Teile der Arbeit zu sehen. Drinnen befindet sich zuerst ein geschlossenes Objekt, ein negatives Raummodell: die Wände, die dieses undurchdringliche Volumen umschreiben, bestehen aus Fotos des Innenraums. Hinzu kommt ein kleineres Drahtmodell als maximale Abstraktion der linearen Form des architektonischen Raumes. In der Mitte steht ein offenes Podest: der Betrachter kann wie in einen Brunnen hineinschauen und sich durch die interne Projektion einer Computer-Animation in den immateriellen synthethischen Raum des Rechners führen lassen.

Plan

FAREED ARMALY

The new Design Center and a section of the Ars Electronica join together under the thematic of "Intelligent Environment".

The other AE sites, Bruckner Haus or the Landesmusem, belong to a history of permanent, purpose-built cultural spaces. Cultural activities occur within fixed rooms, allowing for the changes required for traditional classical forms. The rooms permanent walls are drawn out from the logic of the architectural plan.

Cultural spaces indicate a program that implies a connection to a public, the social life of the city, and symbolic to the city plan. There are links between, for example, a certain sense of what should, or could be, experienced as 'culture' within the 'interior' space, and that expressed via the architectural 'exterior', as orientated within the overall city plan.

The DC is another type of structure. Its plan and program develop out from other orientations. As with all large halls, an important originating point is the intention to 'supply' an environment. Particularly with the scale of such a space, the approach to planning concerns the unpredictable qualities related to: nature; covering the largest volume of unobstructed area; achieving a space with the most possible uniform units; a strict operating economy; large volumes of attending visitors.

The DC space visually, and conceptually, offers a new take on old situations. This is in relation to several histories of building solutions, of a type where engineering technologies responded to their associated concerns (problems relative to statics, etc) which resulted in experiences thought of as belonging to the 'higher' concerns of architecture.

The space should not be judged as simply a success in terms of 'building from the ground up' (and hopefully sustaining that) but in terms of what is indicated as 'environment'. This is evident in the kind of approach concerning the interweave of various engineering technologies. A more apt description of the DC might be a 'user- area' defined by two 'interfaces': the floor and the ceiling.

„Intelligent Environment" – eine Ausstellung von Ars Electronica im neuen Design Center (DC).

Die anderen Veranstaltungsorte von Ars Electronica, das Brucknerhaus und das Landesmuseum, sind permanente, zu einem bestimmten Zweck gebaute Kulturstätten. Kulturelle Aktivitäten finden in Räumen statt, die nur so viel Veränderung zulassen, wie für die traditionellen klassischen Kunstformen notwendig ist. Die feststehenden Wände der Räume folgen der Logik des architektonischen Entwurfs.

Kulturstätten sind programmatisch mit einer Öffentlichkeit, dem gesellschaftlichen Leben einer Stadt, und symbolisch mit der Anlage der Stadt verbunden. So besteht zum Beispiel ein Zusammenhang zwischen dem Gefühl dafür, was als „Kultur" in einem „Innenraum" erlebt werden sollte oder könnte, und dem, was im „Äußeren" der Architektur, in seiner Orientierung innerhalb der Anlage der Stadt, Ausdruck findet.

Das DC ist eine andere Art von Bauwerk. Entwurf und Programm folgen anderen Ausrichtungen. Wie bei allen großen Hallen ist ein wichtiger Ausgangspunkt das Ziel, ein Environment „liefern2 zu wollen.

Vor allem bei Räumen dieser Größenordnung wird der Entwurf bestimmt von Aspekten wie: unvorhersehbare natürliche Einflüsse; die Überspannung eines größtmöglichen durchgehenden Volumens; die Schaffung eines Raumes mit möglichst gleichen Einheiten; eine strikte betriebliche Ökonomie; große Besucherzahlen.

Sowohl visuell als auch konzeptionell bietet das DC eine neue Auseinandersetzung mit alten Situationen an. Das heißt, mit architektonischen Lösungen im Laufe der Geschichte, bei denen die sogenannten „höheren" Anliegen der Architektur die technische Planung bestimmten (z.B. in Hinblick auf die Statik).

Ein Raum sollte nicht nur danach beurteilt werden, wie gelungen er als bauliche Konstruktion ist, sondern auch in bezug auf das, was als „Environment" angedeutet wird. Ein Ansatz, der verschiedene Technologien miteinander verwebt, macht das erkennbar. Das DC ließe sich vielleicht treffender als „user area" bezeichnen, die von zwei „interfaces" definiert ist: dem Boden und der Decke.

Interface 1) Boden

Die Ausstellung von Ars Electronica findet im Obergeschoß des DC statt. Der Fußboden ist für den Raum emblematisch. Seine scheinbare Massivität entpuppt sich bald als nichts anderes als eine Verkleidung, durch die man Zugang zu verschiedenen Versorgungssystemen hat. Die Oberfläche besteht aus 50 x 50 cm großen Platten. Jede Platte kann abgenommen werden und bietet damit einen Zugriff auf die darunterliegende Ebene. Die einzelnen Platten können auch gegeneinander, z.B. gegen Platten mit Druckluftanschlüssen, ausgetauscht werden. Im darunterliegenden Raum werden Lüftungs- und Wasserleitungen geführt, die an die Hauptversorgungsstränge angeschlossen sind. Dieses Environment muß aber nicht nur mit Luft, Wasser und Strom versorgt werden, sondern auch mit Leitungen für Telefon, Computer und Fernsehen.

Interface 2) Deckenkonstruktion

Das Glasdach überspannt das gesamte Gebäude und scheint auf beiden Seiten fast den Boden zu berühren. Der Schlüssel zum DC ist der Umgang mit seinem immateriellsten Element – dem Licht. Hier ist ein Schritt vollzogen von reiner Transparenz, die Hinaussehen ermöglicht und Licht einläßt, hin zum „Interface". Die Decke ist in diesem Sinne kein „Fenster", sondern eine Anordnung von Filtern. Zwischen Isoliergläsern befinden sich Lamellen, oder Lichtraster. Die Position jedes einzelnen Glaspaneels und der darin integrierten Lichtraster in der bogenförmigen Dachkonstruktion ist im Hinblick auf den Einfallswinkel der Sonne genau berechnet.

Wie in der Optik wird das einfallende Licht gefiltert und wie etwas Materielles geformt. Das Bauwerk – ein überdachter Raum – steht hier nicht mehr für die Leistung, große Spannweiten zu realisieren, sondern für den Punkt, an dem Licht gefiltert, verändert und geformt wird wie etwas Materielles.

Es ist eine über Interfaces erfahrbare Welt. Das Gefühl für den Raum, unsere Beziehung im Raum und zur Welt draußen, entsteht nicht nur aus dem „Da-Sein", sondern auch aus dem Environment, der Filterung und Konvertierung der Übertragung verschiedener Elemente, nicht zuletzt des Lichts.

Der Umgang mit Licht ist historisch gesehen ein grundlegendes strukturierendes Element. Das nähert das DC an Räume wie gotische Kathedralen, Ausstellungsglaspaläste, Kinos an. Darüber hinaus stellt es aber als Environment auch eine Verbindung her zu einem Raumgefühl, wie wir es vom Fernsehen, vom Computerbildschirm und der kommenden neuen virtuellen Technologie kennen. Immer geht es dabei um den Fluß und die Formation einer Öffentlichkeit.

Das Decken-Interface läßt auch noch auf eine andere Art die Funktion der Halle als Ort des Handels anklingen. In der Dämmerung bekommt das Glas durch die integrierten Lichtraster noch eine zusätzliche Funktion – die eines Spiegels.

Interface 1) the floor

The Ars Electronica exhibition takes place on the upper floor of the DC. The floor is emblematic for the space. Its apparent solidity is quickly revealed to be nothing more than a cover through which to plug-in to 'services'. Its surface is comprised of 50 cm square tiles. Each can be pulled away to reveal an area underneath. Tiles an be exchanged with variations containing, for example, air-ducts. The space underneath awaits the positioning of channels for bringing and removing air, water, and plugs to be brought in from the main support lines. The necessary supplies for the environment are not just air, water or electricity, but now as well telephone, computer, and TV.

Interface 2) Ceiling structure

The glass ceiling spans the entire center, appearing to touch ground on both sides. The key of the DC is the handling of its most immaterial quality – the light. The change occurs here, from transparence, the ability for a view outwards, recieving light, to 'interface'. There is in fact no 'window' here, but a set of filters. The clear glass actually contains 'lamellen'. Each glass piece, and the angle of the 'lamellen' within, is individually calculated for its position in the arch, relative to the calculated positions of the sun.

As is the case with the physics of optics, such a filtering brings the light in, and shapes it as if to appear as material. Here the historical sign of engineering – a spanned area – no longer just stands for the ability to surpass vast areas, but as the point where occurs a filtering, converting, and shaping of light as itself a material property.

It presents an interfaced world. The sense of a space, our relations within and to the world outside, is achieved not only by physically 'being there', but as well, via the environment of filtering and converting the transmissions of various properties, not the least of which is indicated through light.

The use of light is a main historical structural orientation. Matching this to the DC offers an analogy to such spaces as gothic cathedrals, glass exhibition palaces, cinemas. But as environment, it also establishes a link to the sense of space we are accustomed to as TV, computer monitors and the coming new virtual technology. All of which are involved in respect of the flows and formations of large publics.

In terms of the publics, the hall ceiling interface has one other subtle reminder of the hall's function as a site for commerce. As day light shifts towards dusk, the properties of the lamellan introduce to the transparent glass, to the quality of light, one other – the reflection of the mirror.

Exhibition design as 'reflection'

As with all halls, the DC comes with no set interior walls. A complete interior was required for this Ars Electronica section. In terms of exhibition planning, having fixed interior walls means there are some discussions already concluded. The more concrete the walls, the more final the discussion.

Temporary architecture is an element every large hall requires. The walls stand as a line drawn between what is necessary for two kinds of spaces – the DC, and that of the displays. The DC's permanent rules (access for wheelchairs and fire exits, only free-standing structures, etc.) relate to what shouldn't occur in virtue of the large scale public structure. The rules for the temporary installations begin relative to what should occur in one designated room.

Any contemporary reference to 'environment' includes within it the flow of the media. It may be helpful to consider exhibition architecture as analogous to media. It has more in common with that, than the responsibilities of architecture. Posters, labels and guides are literally as equal in importance – and weight – as the walls. What is conveyed is a sense on display, a space, to the public as a flow of information.

In relation to such exhibitions, the mass media attempts to structure the public around information related to technology advancements. The mass media always works on building anticipation. This happens in particular with new future technologies, the type which profess the ability to allow one to fully experience another world. Virtual Reality is one example.

As such, it is easy to make the leap of faith, and assume that installations involving displays, environments, new technologies, etc., require less of what a traditional room delivers – i.e. heavy, load-bearing, structures – and more a 'framing' which indicates 'space'. In theory, decisions would be more related to indicating an area, for example, required in terms of viewing, or trying out, a head-mounted display set-up.

The development for the exhibition design pointed out that often the opposite was true. Planning the temporary structures, in respect of the

Ausstellungsdesign als „Reflexion"

Wie bei allen Hallen gibt es auch beim DC keine fixen Innenwände. Dieser Teil von Ars Electronica verlangte allerdings komplette Raumkörper. In bezug auf die Planung einer Ausstellung schließen fixe Innenwände bestimmte Auseinandersetzungen bereits aus. Je konkreter die Wände, desto endgültiger die Auseinandersetzung. Sehr große Hallen verlangen eine temporäre Architektur. Die Wände stehen zwischen zweierlei räumlichen Erfordernissen – den Erfordernissen des DC und jenen der Ausstellung. Die permanenten Richtlinien des DC (Zufahrt für Rollstühle, Notausgänge, nur freistehende Konstruktionen, etc.) beziehen sich auf den großen öffentlichen Raum, in dem bestimmte Dinge nicht geschehen dürfen. Die Richtlinien temporärer Installationen zielen auf das ab, was in einem bestimmten Raum passieren soll.

Jede Betrachtung eines „Environments" schließt heute den Fluß der Medien mit ein. Ausstellungsarchitektur sollte deshalb vielleicht als analog zu den Medien betrachtet werden, mit denen sie mehr verbindet als mit den herkömmlichen Verantwortlichkeiten der Architektur. Poster, Schilder und Führer haben buchstäblich dasselbe Gewicht wie die Wände. Was dem Publikum als Informationsfluß mitgeteilt wird, ist ein Gefühl für die Ausstellung, für den Raum.

Die Massenmedien versuchen, die Öffentlichkeit um Information zu strukturieren, die mit dem technologischen Fortschritt zu tun hat. Sie bauen immer Erwartungen auf. Vor allem geschieht das bei neuen, zukunftsweisenden Technologien, die versichern, uns eine andere Welt erfahren zu lassen. Die Virtuelle Realität ist nur ein Beispiel.

In diesem Sinne ist es verständlich, daß Installationen mit Displays, Environments, neuen Technologien, etc. nicht so sehr das brauchen, was ein traditioneller Raum anbietet – d.h. schwere, tragfähige Konstruktionen -, sondern vielmehr „Rahmen", die „Raum" andeuten. Theoretisch drehen sich Entscheidungen mehr darum, wie zum Beispiel eine Fläche, die man zum Betrachten oder Ausprobieren einer bestimmten Präsentation braucht, angedeutet werden kann.

Während der Planungsarbeiten wurde klar, daß oft das Gegenteil zutraf. Die Planung temporärer Konstruktionen in Übereinstimmung mit den Erfordernissen der Installationen bedeutete oftmals, daß man sich komplexere Synthesen überlegen mußte, temporäre Konstruktionen, die genau die Eigenschaften nachahmten, die man mit permanenten Räumen assoziiert – tragfähige Konstruktionen, überdachte Räume, Dunkelheit, Stille und so weiter.

Das ist das Gegenteil von dem, was man bei einer großen Halle als Innenarchitektur erwarten würde. Das Ausstellungsdesign, die Linie, die zwischen der Architektur des DC und den Installationen zu ziehen war, entwickelte sich zu einem „Nexus" verschiedener Philosophien, die sich nicht immer ganz harmonisch, aber doch im

Zweiklang zueinander verhielten: technische Lösungen, philosophische Standpunkte, Weltauffassungen – und manchmal einfach der Unterschied zwischen „Raum" und „Raum = Zimmer".

Einige technische Debatten über Installationen zukünftiger Technologien mögen mitunter unwesentlich scheinen, doch stehen sie für allgemeinere philosophische Haltungen.

Es geht darum, welche Beziehungen innerhalb der Parameter dieser Welt ein Individuum erfahren soll. Es ist zum Beispiel interessant, wie man an die Frage herangeht, ob die Massivität und Attraktion der Technik, von Maschinen, Projektoren, Computern, etc. „negiert" werden soll oder nicht.

Der Rahmen, der gesetzt werden muß, damit die Arbeiten „funktionieren" können, ist ein Environment, das, wie es scheint, sich immer noch mit traditionellen Fragen auseinandersetzen muß, wie etwa mit dem Gewicht, das eine Ansammlung sich bewegender Menschen darstellt, mit Fragen der Statik, der Schwerkraft, etc. In diesem Sinne ist der Ausstellungsort auch ein exemplarischer zeitgenössischer Kommentar.

Der Entwurf verbindet die Ausstellung mit der Architektur und den Designstärken des Ortes. Arbeiten, die für sich alleine stehen können, einen „Raum" ignorieren, negieren oder nicht brauchen, wurden mit Blick auf jene Installationen positioniert, die bestimmte „räumliche" Elemente brauchen. Diese sind zusammengruppiert, teilen sich tragfähige Wände und schaffen so den größtmöglichen Raum auf einer begrenzten Fläche, so daß von außen gesehen ein einheitlicher strukturierter „Innenraum" entsteht. Durch das Gesamtdesign, die Maßstabsverhältnisse und die symmetrische Positionierung werden Beziehungen aufgenommen, die in der Logik des Plans des DC ausgedrückt sind.

Mitarbeit: Ulrike Kremeier

Foto: Horst Jaritz

installation's demands, often meant having to consider somewhat a more complex synthesis, a temporary structure mimicking exactly those qualities associated with permanent rooms – load bearing structures, spanned ceilings, darkness, silence, and so on.

This is the opposite of what any large hall sets up to anticipate occurring as interior architecture. The exhibition design, the line drawn between the DC's architecture and the installations within, developed as a 'nexus' of various philosophies operating not always in tune but certainly in tandem with each other. These concern engineering technology, philosophical outlooks, worldviews – and sometimes simply the difference between a 'room' and 'space'.

Regardless of whether God or the devil is in the details, the level of some of the technical discussions concerning future technology installations may appear insignificant at points, but in fact, these reveal and stand for an overall philosophical outlook.

It concerns what kind of relations are being set up, for an individual to experience, within the paramters of this world. It is informative to note how a concept design approaches, for example, whether or not to 'deny' the massivity and attraction of the engineering, the machines, projectors, computers, etc.

The framework which must be established, to allow the works to 'work', is an environment, that appears still has to begin with the traditional questions taking into account the weight and force of an accumulation of moving bodies, statics, gravity etc – the traditions of physical engineering. As such, the host site stands out as an exemplary contemporary comment on this plain truth about forces.

The design works to link the exhibition to the architecture and design strengths offered by the site. Works which can stand alone, ignore, deny, or be free of a 'room' were situated in respect of those various individual installations that required certain specific 'room' interior factors. These benefitted by being grouped together so as to share walls for greater support, bring about the most space possible within a limited area, and as allow from an exterior vantage point, one apparent overall structured 'interior'. The structure's overall design, scale, symmetrical placement is intended to link up to relations expressed in the logic of the Design Center plan.

BAR CODE HOTEL

PERRY HOBERMAN

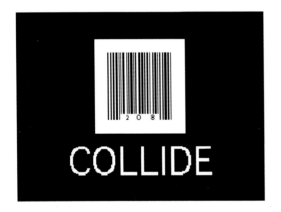

Das *Bar Code Hotel* ist ein interaktives Environment für mehrere Teilnehmer (oder Gäste). Ein ganzer Raum wird mit Strichcode-Symbolen ausgekleidet, wobei jede Fläche zu einer reaktiven Membran werden kann, zu einem Interface, das gleichzeitig von mehreren Personen benutzt werden kann, um eine virtuelle computergenerierte Echtzeit-3-D-Welt zu steuern und mit ihr in Interaktion zu treten.

Jeder Gast, der ins *Bar Code Hotel* kommt, erhält einen Strichcode-Stab, einen leichten Stift, mit dem sich die gedruckte Strichcode-Information sofort in das Computersystem einscannen läßt. Da jeder Stab vom System als eigenes Eingabegerät identifiziert wird, besitzt jeder Gast innerhalb der computergenerierten Welt eine eigene Identität und Persönlichkeit. Und da das Interface der Raum selbst ist, können die Gäste nicht nur mit der computergenerierten Welt in Interaktion treten, sondern auch miteinander. Die Strichcode-Technologie bietet praktisch unendlich viele wartungsfreundliche Sensorgeräte an (die einzige Beschränkung ist der zur Verfügung stehende physische Raum), mit denen sich jeder Quadratzentimeter der Oberfläche des Raumes in die virtuelle Welt des Computers übertragen läßt.

Das virtuelle Environment besteht aus mehreren computergenerierten Objekten, die jeweils einem bestimmten Gast entsprechen. Diese Objekte entstehen, indem einzelne Strichcodes, die auf weiße, im ganzen Raum verteilte, Würfel aufgedruckt sind, eingescannt werden. Sind die Objekte einmal geschaffen, führen sie eine halbautonome Existenz und sind nur zum Teil von den menschlichen Akteuren steuerbar. Sie reagieren auch auf andere Objekte und auf ihre Umgebung. Bei ihren Aktionen und Interaktionen produzieren sie verschiedene Töne. Sie haben ihr eigenes Verhalten, ihre eigene Persönlichkeit; sie haben auch ihre eigene Lebenszeit (in der Größenordnung von einigen Minuten); sie altern und sterben (letztendlich).

Das *Bar Code Hotel* kann eine beliebige Zahl von Gästen aufnehmen, je nachdem, wieviele Strichcode-Stäbe vorhanden sind (was wiederum von der jeweils installierten Ausstattung abhängt). Zur Zeit können im Hotel problemlos bis zu sechs Gäste untergebracht werden.

Immer, wenn ein Gast einen Strichcode einliest, wird ein Kontakt zwischen ihm und seinem Objekt hergestellt. Zwischen den Momenten, in denen sie mit einem Menschen in Kontakt treten, sind die Objekte allerdings auf sich selbst gestellt. Das ermöglicht eine ganze Reihe denkbarer Interaktionsarten. Gäste können, wenn sie wollen, mit ihrem Objekt immer Kontakt halten, indem sie praktisch ununterbrochen Befehle einscannen. Oder sie können ihren Einfluß reduzieren, beobachten, was passiert, und nur gelegentlich ihrem Objekt „beratend" zur Seite stehen.

Alle Strichcodes können zu jeder Zeit eingescannt werden. Jeder Strichcode ist (verbal oder graphisch) ge-

Bar Code Hotel is an interactive environment for multiple participants (or guests). By covering an entire room with printed bar code symbols, an installation is created in which every surface can become a responsive membrane, making up an immersive interface that can be used simultaneously by a number of people to control and respond to a projected real-time computer-generated three-dimensional world.

Each guest who checks into the *Bar Code Hotel* is given a bar code wand, a lightweight pen with the ability to scan and transmit printed bar code information instantaneously into the computer system. Because each wand can be distinguished by the system as a separate input device, each guest can have their own consistent identity and personality in the computer-generated world. And since the interface is the room itself, guests can interact not only with the computer-generated world, but with each other as well. Bar code technology provides a virtually unlimited series of low-maintenance sensing devices (constrained only by available physical space), mapping every square inch of the room's surface into the virtual realm of the computer.

The projected environment consists of a number of computer-generated objects, each one corresponding to a different guest. These objects are brought into being by scanning unique bar codes that are printed on white cubes that are dispersed throughout the room. Once brought into existence, objects exist as semi-autonomous agents that are only partially under the control of their human collaborators. They also respond to other objects, and to their environment. They emit a variety of sounds in the course of their actions and interactions. They have their own behavior and personality; they have their own life span (on the order of a few minutes); they age and (eventually) die.

Bar Code Hotel is designed to accomodate any number of guests, up to the available number of bar code wands (which is dependent on the particular configuration installed). Currently, the Hotel can easily handle between one and six guests at a time. Each time a guest scans a bar code, contact is re-established between that guest and their object. However, between these moments of human contact, objects are on their own. This allows for a number of possible styles of interaction. Guests can choose to stay in constant touch with their object, scanning in directives almost continuously. Or they may decide to exert a more remote influence, watching to see what happens, occasionally offering a bit of "advice". Guests can scan any bar code within reach at any time. Each bar code is labeled (verbally or graphically), letting the user know what action will result.

The objects in *Bar Code Hotel* are based on a variety of familiar and inanimate things from everyday experience: eyeglasses, hats, suitcases, paperclips, boots, and so on. None of them are based on living creatures; their status as characters (and as surrogates for the user) is tentative, and depends totally upon their movement and interaction. At times they can organize themselves into a sort of visual sentence, an unstable and incoherent rebus.

Objects can interact with each other in a variety of ways, ranging from friendly to devious to downright nasty. They can form and break alliances. Together they make up an anarchic but functioning ecosystem. Depending on their behavior, personality and interactive "style", these objects can at various times be thought of in a number of different ways. An object can become an agent, a double, a tool, a costume, a ghost, a slave, a nemesis, a politician, a pawn, a relative, an alien. Perhaps the best analogy is that of an exuberant and misbehaving pet.

Bar codes can be scanned to modify objects' behavior, movement and location. Objects can expand and contract; they can *breathe, tremble, jitter* or *bounce*. Certain bar code commands describe movement patterns, such as *drift* (move slowly while randomly changing direction), *dodge* (move quickly with sudden unpredictable changes) and *wallflower* (hide in the nearest corner). Other bar code commands describe relations between two objects: *chase* (pursue nearest object), *avoid* (stay as far away as possible from all other objects), *punch* (collide with the nearest object) and *merge* (occupy the same space as the nearest object). Of course, the result of scanning any particular bar code will vary, based on all objects current behavior and location. Many bar code commands cause temporary appendages to grow out of objects. These appendages amplify and define various behaviors. Particularly aggressive objects often grow spikes, for example.

Each object develops different capabilities and characteristics, depending on factors like age, size and history. For instance, younger objects tend to respond quickly to bar code scans; as they age, they become more and more sluggish. Older objects begin to malfunction, short-circuiting and flickering. Finally, each object dies, entering briefly into an ghostly afterlife. (This process can be accelerated by scanning suicide.) After each object departs, a new object can be initiated.

Besides controlling objects, certain bar codes affect and modify the environment in which the

kennzeichnet, so daß der Teilnehmer weiß, welche Aktion er auslöst.

Die Objekte im *Bar Code Hotel* basieren auf unbelebten Dingen aus dem täglichen Leben: Brillen, Hüte, Koffer, Büroklammern, Schuhe, etc. Kein Objekt basiert auf einem Lebewesen. Ihr Status als Person (und als Ersatz für den Teilnehmer) ist tentativ und hängt ganz von ihrer Bewegung und Interaktion ab. So können sie sich zum Beispiel auch zu einer Art visuellem Satz organisieren, einem instabilen und nicht kohärenten Bilderrätsel.

Die Objekte können miteinander auf verschiedene Art in Interaktion treten, freundlich, listig oder sogar böse. Sie können Allianzen bilden und wieder auflösen. Gemeinsam bilden sie ein anarchisches, aber funktionierendes Ökosystem.

Je nach Verhalten, Persönlichkeit und „Interaktionsstil" kann man sich diese Objekte zu verschiedenen Zeiten in verschiedenen Rollen denken. Ein Objekt kann Agent sein, Doppelgänger, Werkzeug, Kostüm, Geist, Sklave, Nemesis, Politiker, Pfand, Verwandter, Fremder. Die beste Analogie ist vielleicht die eines übermütigen, schlecht erzogenen Haustieres.

Durch das Einscannen der Strichcodes kann man das Verhalten, die Bewegungen und die Position der Objekte verändern. Objekte können sich ausdehnen und zusammenziehen; sie können *atmen, zittern, bibbern* oder *springen*. Bestimmte Strichcode-Befehle beziehen sich auf Bewegungsmuster, z.B. *Treiben* (sich langsam bewegen und dabei wahllos die Richtung ändern), *Sprung* (sich schnell bewegen, mit plötzlichen unvorhersehbaren Richtungsänderungen) und *Mauerblümchen* (sich in die nächstgelegene Ecke zurückziehen). Andere beeinflussen die Beziehung zwischen zwei Objekten: *Jagen* (das nächstbefindliche Objekt verfolgen), *Vermeiden* (so weit weg wie möglich von allen anderen Objekten bleiben), *Schlagen* (mit dem nächstbefindlichen Objekt zusammenstoßen) und *Vereinigen* (denselben Raum einnehmen wie das nächstbefindliche Objekt). Natürlich ist das Ergebnis, wenn man einen bestimmten Strichcode einliest, nicht immer dasselbe und hängt vom Verhalten und der Position aller Objekte zu diesem Zeitpunkt ab.

Auf manche Strichcode-Befehle hin wachsen vorübergehend Anhängsel aus den Objekten heraus. Diese Anhängsel verstärken und definieren bestimmte Verhaltensmuster. So wachsen aus besonders aggressiven Objekten oft Stacheln heraus.

Jedes Objekt entwickelt unterschiedliche Fähigkeiten und Eigenschaften, je nach Alter, Größe, Erfahrung usw. Zum Beispiel reagieren jüngere Objekte meist schneller auf Strichcode-Befehle; je älter sie werden, desto träger werden sie. Ältere Objekte versagen oft, flimmern oder haben einen Kurzschluß. Irgendwann stirbt jedes Objekt, wobei es nach seinem Tod noch kurz als Geist weiterlebt. (Dieser Prozeß kann beschleunigt werden, indem man Selbstmord einscannt.) Nach dem Verschwinden eines Objektes kann ein neues Objekt initiiert werden. Mit einigen Strichcodes kann man nicht nur die Objekte steuern, sondern auch die Umgebung, in der sie existieren, beeinflussen und verändern. Der Standpunkt der Computerprojektion kann verlagert werden. Als Hintergrund können verschiedene Räume und Landschaften gewählt werden. Es können kurze Erdbeben kreiert werden (die die Objekte in einen Zustand äußerster Orientierungslosigkeit stürzen).

Nachdem jeder Strichcode zu jeder Zeit eingescannt

objects exist. The point of view of the computer projection can be shifted. Settings can be switched between various rooms and landscapes. Brief earthquakes can be created (leaving all objects in a state of utter disorientation).

Since any bar code can be scanned at any time, the narrative logic of *Bar Code Hotel* is strictly dependent on the decisions and whims of its guests. It can be played like a game without rules, or like a musical ensemble. It can seem to be a

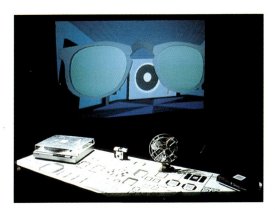

werden kann, hängt die narrative Logik des *Bar Code Hotels* strikt von den Entscheidungen und Launen seiner Gäste ab. Es kann wie ein Spiel ohne Spielregeln gespielt werden oder wie ein musikalisches Ensemble. Es kann aussehen wie ein langsamer, eleganter Tanz oder wie eine Slapstick-Komödie. Und weil die Aktivitäten des *Bar Code Hotels* sowohl von seinen jeweiligen Gästen abhängen als auch vom autonomen Verhalten der verschiedenen Objekte, sind unendlich viele unvorhersehbare und dynamische Szenarien möglich.

slow and graceful dance, or a slapstick comedy. And because the activities of *Bar Code Hotel* are affected both by its changing guests and by the autonomous behavior of its various objects, the potential exists for the manifestation of a vast number of unpredictable and dynamic scenarios.

Bar Code Hotel was developed as part of the Art and Virtual Environments Project at the Banff Centre for the Arts. The Project was sponsored by the Computer Applications and Research Program at the Banff Centre, which was funded by The Department of Canadian Heritage and CITI (Centre for Information Technologies Innovation). The Banff Centre also received support from: Silicon Graphics Inc., Alias Research, The Computer Science Department of the University of Alberta, Apple Canada, Intel, and AutoDesk Inc.

Graphics Programming & System Design: John Harrison, Glen Fraser, Graham Lindgren / Sound Design: Dorota Blaszczak, Glen Fraser / Graphics Design: Cathy McGinnis / Poject Director: Douglas MacLeod / Special thanks to Steve Gibson, Ron Kuivila, Doug Smith, Sylvie Gilbert, Daina Augaitis, Tim Westbury, Mimmo Maiolo & Angela Wyman.

SPATIAL LOCATIONS, VERSION III

HERMEN MAAT / RON MILTENBURG

"Tout le malheur des hommes vient d'une seule chose, qui est ne savoir pas demeurer en repos, dans une chambre."
Blaise Pascal, "Pensées"

I. Imagine a world in which Europe hasn't a clue as to the direction in which it will develop. A world in which professors tell us that Islam is a religion to be respected, as good for Muslims as Christianity is for Christians. A world in which a bald head with a beard is just as acceptable as a longhaired head with a clean shaved chin. A world in which new technologies announce themselves and evoke dimensions yet to be thought of. A world in which the consequences of these developments for the constellation of states, taxes and moralities can only be second guessed. A world in which scientific data are as good as dogmatic teachings or the opinions in the streets. Man looks into the abyss of civilisation.

That world existed. Back at the beginning of the 17th century. It is the world of Blaise Pascal and his contemporaries. Imagine the horror generated by Montaigne a few decades before: Que

„Tout le malheur des hommes vient d'une seule chose, qui est ne savoir pas demeurer en repos, dans une chambre."
Blaise Pascal, „Pensées"

I. Stellen Sie sich eine Welt vor, in der Europa keine Ahnung hat, in welche Richtung es sich entwickeln wird. Eine Welt, in der die Professoren uns sagen, daß der Islam eine Religion ist, die man respektieren muß und die für Moslems ebenso gut ist wie das Christentum für die Christen. Eine Welt, in der eine Glatze und ein Bart ebenso akzeptabel sind wie langes Haar und ein glattrasiertes Kinn. Eine Welt, in der sich neue Technologien ankündigen und Dimensionen heraufbeschwören, die wir uns noch gar nicht vorstellen können. Eine Welt, in der die Folgen dieser Entwicklungen für die Konstellation von Staaten, Steuern und Moralvorstellungen sich nur erraten lassen. Eine Welt, in der wissenschaftliche Daten von gleicher Bedeutung sind wie die dogmatischen Lehren oder die Meinungen der Leute von der Straße. Der Mensch blickt in den Abgrund der Zivilisation.

Eine solche Welt existierte schon einmal. In der Vergangenheit, zu Beginn des 17. Jahrhunderts. Es ist die Welt von Blaise Pascal und seinen Zeitgenossen. Stellen Sie sich das

ROCK AND ROLL,
until the fifties the privilege of the elderly and disabled, gained full swing as the chair gave up its categorical functions in the sixties. Present youth and fitness culture have the old rocking chair and wheelchair to thank for its overwhelming success.

HOWEVER, ITS STRICT I
OR ITS TONGUE-IN-CHEEK INDIFFERENCE
IS REMINISCENT OF
HERMEN

The chair most lively of all, of course, is the electric chair, ironically designed to end life. Equally paradoxical is the comfort provided by this piece of furniture. **THE ROUND TABLE ACTUALLY** Both the seated convict and **TODAY IN ROUND TABLE CONFERENCES** the upright bystander are shocked. Each in his own way. **HIS TABLE IS SURELY T** But, as the villain's body jolts, **AS THE TALE OF THE KNIGHTS WAS CONSTRU** the mind of the righteous is put to rest. **WE HAVE NO WAY OF KNOWING W**

Erschrecken vor, das Montaigne einige Jahrzehnte vorher verursacht hatte: Que sais je? Was weiß ich denn? Nichts! Erschreckender als ein ständiger Strom von Desinformation im Fernsehen, in Zeitschriften oder über Internet. Und der blaue Himmel, nicht länger mehr ein göttliches Firmament über ihnen. Nichts als eine optische Täuschung, die sich auflöst und den Blick auf einen Strudel von anderen Welten freigibt. Weniger greifbar als Cyberspace. Eine schreckliche Auswahl an Möglichkeiten eröffnet sich.

Anfangs hält sich Pascal an die wechselnden Launen, Betrachtungsweisen und Modelle der Avantgarden avant la lettre im sehr weltlichen Paris. Gott weiß, was er geschluckt, geschnupft oder geraucht hat, aber eines Nachts hatte er eine unheimliche Begegnung der dritten Art und wurde dadurch zu einem eifrigen Apologeten des Christentums, zu dessen Unterminierung er als Mathematiker, Physiker, Philosoph und Mensch unbeabsichtigt beigetragen hatte.

Pascal, der Autor der „Pensées", umging die Herausforderung des Essayisten Montaigne.

II. Es gibt Essays und es gibt Pensées. Der Essayist lockert seinen Griff auf die Phänomene, wenn er sie einmal begriffen hat. Der Essayist und die freigesetzten Phänomene stürzen gemeinsam auf denselben schrecklichen Abgrund zu, vor dem der Autor der Pensées zurückschreckt. Unterwegs verflüchtigen sich manche Begriffe, doch andere klumpten sich zusammen, nehmen neues Material auf oder verursachen eine Lawine. Der Essayist wird möglicherweise

sais je? What do I know? Nothing! More terrifying than an incessant stream of disinformation on TV, in magazines, or on Internet. And the blue sky, no longer a divine roof over their heads. Nothing but an optical illusion, evaporating, and disclosing a vertigo of universes. Less tangible than cyberspace. A hideous array of maybes opens. At first, Pascal swings with the changing moods, modes and models of the avantgardes avant la lettre in mundane Paris. God knows what he swallowed, snorted or smoked, but one night he had a close encounter of the third kind and consequently became a zealous apologist for the Christianity he unwittingly had helped to undermine as a mathematician, physicist, philosopher and as a living being.

Pensée-writer Pascal side-stepped the challenge of Essay-writer Montaigne.

II. There are Essays and there are Pensées. The Essayist loosens his grip on the phenomena once apprehended. Together, the Essayist and the unleashed phenomena, they hurl themselves downhill towards the same hideous abyss from which the Pensée-writer shies away. En route some apprehensions evaporate, others dab together or pick up new material or cause an avalanche. The Essayist might be snowed under or appear as the Horrible Snowman. In both

ROLE IN TABLE TENNIS,
POOLROOM HUSTLERS FOR THAT MATTER,
AS A MERE OBJECT.

CONTEMPORARY INVENTIONS SUCH AS WATER MATTRESSES AND VIBRATION DEVICES IT HAS BECOME AN ACTIVE THIRD PARTY IN BOTH LOVE AND LIFE.

? CUT IT OUT! THIS IS AN ART SHOW

WE ARE BORN IN BEDS AND EVENTUALLY DIE IN THEM. IN BOTH CASES WE HAVE NO CHOICE.
HE PRINCIPLE OF EQUALITY.   THERE IS ALWAYS A BED WITH OUR NAME ON IT.
STORY AT THE COURT OF KING ARTHUR.   BUT WHAT'S IN A NAME: INVOLUNTARY LIFE AND DEATH.
KNOWN IN LITERATURE.   IN-BETWEEN, WE ARE FREE TO PURCHASE ANY BED THAT MATCHES OUR
F SEVERAL LEGENDS FROM DIFFERENT AGES,   PERSONALITY. SHOW ME YOUR BED AND I'LL TELL
IS ROUND TABLE REALLY EXISTED.   WHO YOU ARE. THIS CATALOGUE OFFERS YOU A WIDE VARIETY.

cases we can only follow his traces and find the spot where his Essay threatened to turn Pensée.

III. In the first chapter of Genesis, God created heaven and earth, the elements, the flora, fauna and man. Nomen est omen, so he provided all his creations with names and meaning, and he saw that it was good. The seventh day he created reflective silence, and saw that it wasn't all that good. Here we enter the second chapter, and God tries it all over again. This time in the garden of virtues, in the reality of virtualis. Upon the end of the chapter he leads all animals on the land and all birds in the sky to man, in order to see how man would name them: "Because, as man would name each living being, thus it would be said to be." And man indicated and said: "Parasite wasps, desert rat, dung beetle." And: "Milking cow, draughthorse, honeybee."

In the following chapter, of which I fail to understand the plot, God eventually dissolves the distinction between his first – real – world and his second – virtual – world. He retired in the heavens and said: "Aprés moi le deluge." And so it came to pass. Within two chapters. Man was made to build an ark by specification clearly described. Then, man embarked with his parasite wasp, desert rat, dung beetle, milking cow, draughthorse and honeybee, two of each.

eingeschneit oder erscheint als der sagenhafte, schreckliche Schneemensch. In beiden Fällen können wir nur seiner Spur folgen und den Punkt finden, an dem sein Essay Gefahr lief, zum „Pensée" zu werden.

III. Im ersten Kapitel des Buches Genesis schuf Gott Himmel und Erde, die Elemente, die Pflanzen, die Tiere und den Menschen. Nomen est omen, daher gab er all seinen Schöpfungen Namen und Bedeutung, und er sah, daß es gut war. Am siebten Tag schuf er die nachdenkliche Stille, und er sah, daß sie nicht allzu gut war. Hier beginnt das zweite Kapitel, und Gott versucht es noch einmal von vorne. Diesmal im Garten der Virtualität, in der virtuellen Wirklichkeit. Am Ende des Kapitels führt er alle Tiere des Feldes und alle Vögel des Himmels dem Menschen zu, um zu sehen, wie der Mensch sie benennen würde: „Und wie der Mensch jedes lebendige Wesen benannte, so sollte es heißen." Und der Mensch gab Namen und sagte: „Schlupfwespe, Känguruhratte, Mistkäfer." Und: „Milchkuh, Zugpferd, Honigbiene."

Im folgenden Kapitel, dessen Handlung sich meinem Verständnis entzieht, löst Gott schließlich den Unterschied zwischen seiner ersten – wirklichen – und seiner zweiten – virtuellen – Welt auf. Er zog sich in den Himmel zurück und sagte: „Après mois le deluge."

Und so geschah es. Innerhalb von zwei Kapiteln. Dem Menschen wurde befohlen, nach genauen Angaben eine Arche zu bauen. Dann bestieg der Mensch mit seiner Schlupfwespe, seiner Känguruhratte, seinem Mistkäfer, sei-

You've got nobody but yourself to blame, punk. Now, sit down, or we'll make you. BOY, AM I GLAD TO SEE YOU! ARTH

We're all set to go Warden. Just say when, and we'll fry the sucker. Yessirree! REJOICE LA DID BRI

THERE YOU ARE GUINEVERE, LOVEL ND
Miss, you're not gonna faint, are you? Please, lie down. It'll be over in a jiffy.
WHAT THE BEEB IS THE MATTER WITH YOU?!

By the book, ladies and gentlemen. This room is off limits to the press. Out! YOU, THE

ner Milchkuh, seinem Zugpferd und seiner Honigbiene, – je einem Paar von ihnen – die Arche.

Durch genaues Lesen der Genesis wissen wir um die besondere Identität der Dinge, die den Menschen umgaben, bevor er den Turm zu Babel zu bauen versuchte.

IV. Il y a une table, es gibt ein Bett, there is a chair. Nicht in der Arche, nein. Oder vielleicht doch. Darüber berichtet die Genesis nichts. (Gott benannte den Himmel und die Erde, die Elemente, die Pflanzen und den Menschen. Der Mensch benannte die Tiere des Feldes und die Vögel des Himmels.) Die Gegenstände mußten offenbar für sich selbst sprechen. Doch infolge des Turmbaus zu Babel verfingen sie sich im Netz von vierzig mal vierzig Grammatiken, und ihre Stimmen wurden niemals mehr gehört.

Was für ein Id haben die Dinge wie *Tisch*, *Bett* und *Sessel*? Was ist ihre Identität und wie ist ihre Stellung in der Welt? *Il y a*, ja, aber was für ein *Er* oder *Es* hat der Tisch; es gibt ein Bett, o.k., aber was für ein *Es gibt* das Bett; *there is*, freilich, aber was ist der *undifferenzierte Raum*, der vom Sessel artikuliert wird? Weder die französische noch die deutsche noch die englische Sprache beantworten diese Frage so, wie es die Gegenstände selbst könnten. Um die Mittel zu finden, die Gegenstände wieder sprechen zu lassen, müssen wir unseren Griff lockern: L'entendement, der Begriff, the appréhension.

Als Gott dem Menschen das Buch der Bücher zum dritten Mal entgegenschleuderte, tat er das im elften Kapitel und hemmte damit den freien Fluß der Rede, des Essays. Wir

By closely reading the Pensée Genesis, we know about the specific identity of the things surrounding man before he attempted to build the tower of Babel.

IV. Il y a une table, es gibt ein Bett, there is a chair. Not in the ark, there wasn't. Or maybe there was, Genesis doesn't show a record on that subject. (God named heaven and earth, the elements, the flora and man. Man named the animals on the land and the birds in the sky). The objects, apparently, had to speak for themselves. However, in the aftermath of Babel they got caught in the web of forty times forty syntaxes, and their voices have never been heard again. Of what Id are the entities *table*, *bed* and *chair*? What is their specific identity and how are they in the world? *Il y a*, yes, but what *he* or *it* has the table; *es gibt*, okay, but what *'it' gives* the bed; *there is*, sure, but what is the *undifferentiated space* articulated by the chair? Neither the French, nor the German, nor the English answer these questions like the objects themselves might. In order to find means to make the objects talk again, we've got to loosen our grip: l'entendement, der Begriff, the apprehension.

When God threw the Book at man, the third time, he did so with chapter 11, foreclosing the free flow of speech, of essay. We might have to

?MY KING! SIRE, WELCOME TO THE WONDERFUL WORLD OF THE BED EXPERIENCE. PLEASE TRY THIS ONE FOR SIZE.

BLE. COME ALONG, A SECOND OPINION IS IN ORDER, WOULDN'T YOU SAY, GIVEN THE SUBJECT?

R. YOU RE INNOCENT, AREN'T YOU? A PARTY OF THREE? A TROIS, AS THE FRENCH WOULD SAY. HOW CAN I BE OF SERVICE TO YOU-ALL? BORING, BORING, BORING

TER! AS FOR YOU OTHER COSTUMERS, NO BUSINESS HERE. PLEASE HELP YOURSELF TO A CUP OF COFFEE IN THE OTHER ROOM.

backtrace his tracks. We might even have to regress beyond the first time, when he tried to cover up discrepancies in this Bookkeeping; we might have to re-install the distinction between the real and the virtual.

V. Whether God really exists or not, cannot be determined by reason. There is this fifty-fifty chance. Belief might win you eternal bliss ...if God exists. If he doesn't, you haven't lost anything. Like you wouldn't if you did believe but he didn't exist. On the other hand, if he does exist, and you didn't believe, you have lost. Forever.

This reasoning, known as Pascal's Wager, is characteristic of the Pensée-writer at large. Some 250 years ago Blaise Pascal bet man's spiritual welfare on the assumptions of Christianity. Some 5 years ago Alain Finkielkraut waged man's salvation on the assumptions of Culture. Both, scholars of their times, could not but discover unmarked slopes, yet condemned man to langlaufen. God, Pascal, Finkielkraut and all the other Pensée-writers, became zealous apologists for the preconceived ideas they once unwittingly had endangered by their essayistic nature. Somewhere between the rise and fall of the Pensée, somewhere between God's *Genesis* and Finkielkraut's *Defaite*, Pascal knew: "Les extrèmes se touchent."

müssen seine Spuren vielleicht bis zum Anfang zurückverfolgen. Wir müssen vielleicht sogar noch weiter zurückgehen als bis zum ersten Mal, da er Unstimmigkeiten in dieser Buchführung zu verbergen trachtete; wir müssen vielleicht den Unterschied zwischen dem Wirklichen und dem Virtuellen wiederherstellen.

V. Ob Gott wirklich existiert oder nicht, läßt sich mit dem Verstand nicht feststellen. Die Chancen stehen 50:50. Der Glaube kann vielleicht die ewige Glückseligkeit erringen ... wenn Gott tatsächlich existiert. Wenn er nicht existiert, ist nichts verloren. So wie man nichts verlöre, wenn man an ihn glaubt, er aber nicht existierte. Andererseits jedoch, wenn er existiert und man nicht geglaubt hat, hat man verloren. Für immer.

Diese Argumentation, die als die sogenannte Wette Pascals bekannt ist, ist insgesamt für den Autor der Pensées charakteristisch. Vor etwa 250 Jahren setzte Blaise Pascal das geistige Heil des Menschen auf das Postulat des Christentums. Vor etwa fünf Jahren setzte Alain Finkielkraut die Erlösung des Menschen auf das Postulat der Kultur. Beide konnten als Gelehrte ihrer Zeit nichts als Abhänge ohne markierte Pisten entdecken und verdammten den Menschen dennoch zum Langlaufen. Gott, Pascal, Finkielkraut und all die anderen Pensées-Schreiber wurden zu eifrigen Apologeten der vorgefaßten Ideen, die sie durch ihre essayistische Natur einst unbeabsichtigt gefährdet hatten. Irgendwo zwischen dem Aufstieg und Untergang der Pensées, irgendwo zwischen Gottes *Genesis* und Finkiel-

You'll sit this one out. OUT, gottit? Sitting it OUT...

ARTHA, MY MAN, THAT TAB
IS THAT A GUN

Do us the honors Warden, chair this session, as it were.

DON'T DASH OFF AGAIN L

And then we can all take a rest. In that bed-chair over there.
COME HERE OFTEN?

LET'S INVITE PARCIVAL. G

Chairman, chairlady, chairperson. What a world. Sod off!

TURNING TABLES, MULTIPL

This work of art has been made possible by the FOUNDATION FOR VISUAL ARTS, DESIGN AND ARCHITECTURE – AMSTERDAM.

krauts *Defaite* wußte Pascal: „Les extrêmes se touchent."

VI. Ich werde Zimmer genannt, aber ich weiß nicht, was das bedeutet. Ich habe Zimmerkollegen: einen Sessel, einen Tisch und ein Bett. Ich weiß nicht, was das bedeutet, und auch sie wissen es nicht. Ich habe Ihnen soeben fünf Geschichten erzählt; ich kann nur hoffen, daß ich sie richtig erzählt habe. Dort, wo sie herkommen, gibt es noch viele andere. Jede Faser unseres Seins ist in Geschichten getaucht. Wir hören sie seit vielen Jahrtausenden. Aber wir wissen nicht, was sie bedeuten. So viel haben wir gesammelt, wir sind die einzigen Überreste einer riesigen schwimmenden Vorrichtung. Es müssen damals kritische Zeiten gewesen sein, denn wir wurden kaum angesprochen oder verwendet. All das hat sich im Laufe der Zeitalter geändert. Wenn wir noch einen Besucher bekommen, erkennen wir manchmal die Art, auf die er uns benutzt. Sie wird zu einem Teil unserer Identität, aber wir wissen nicht, was das bedeutet, außer wenn dieser Besucher anwesend ist. Wir mögen das. Wir erzählen unserem Besucher gerne, was er aus uns gemacht hat. Wenn er lange genug bleibt, erwärmen wir uns für den Besucher. Solange er uns benutzt. Sonst bekommen wir schlechte Laune, aber wir wissen nicht, was das bedeutet. Wir haben eine Frage an Sie. Viele Besucher beklagen ihre Existenz und/ oder ihre Identität. Wir hören sie sagen: „Wenn Gott doch bloß ein Zimmer geschaffen hätte, nur ein Zimmer, und Besuchern erlaubt hätte, hineinzuschauen und zu sagen: 'Zeig mir deine Möbel und ich sage dir, was du bist.'" Was bedeutet das?!

VI. My name is Room, but I don't know what that means. I have roommates: a chair, a table and a bed. I don't know what that means, and neither do they. I've just told you five stories; I can only hope that I related them accurately. There are plenty more where they come from. Each and every fiber of us is drenched in stories. Many a millenium have we heard them. But we don't know what they mean. So much we have gathered, we are the only remains of a huge floating device. It must have been critical times, back then, because we were hardly addressed or used. That all changed over the ages. When we get yet another visitor, sometimes we recognize the way he puts us to use. That becomes part of our identity, but we don't know what that means, unless that visitor is present. We like that. We like to tell our visitor what he made of us. If he stays long enough, we warm up to the visitor. As long as he uses us. Otherwise we get ill-tempered, but we don't know what that means. We have a question for you. A lot of visitors lament their existence and/or their identity. We hear them say: "If God only had created a room, nothing but a room, and allowed visitors to peep in and say 'Show me your furniture and I´ll tell you what you are." What does that mean?!

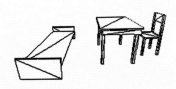

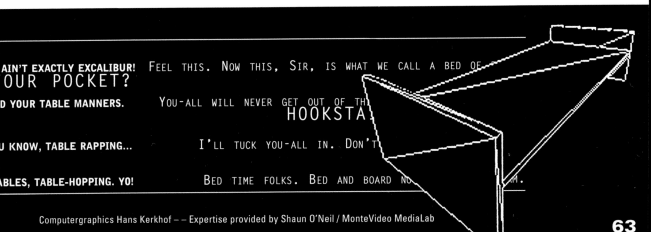

Computergraphics Hans Kerkhof — — Expertise provided by Shaun O'Neil / MonteVideo MediaLab

ARCHITEXTURE
computer-generated pneumatic biogrids

SUPREME PARTICLES

Goal

The observer will be confronted with a variable pneumatic-acoustic screen through the architectural + technological arrangement. Both visual and acoustic perception will be addressed. The screen itself behaves as if it were intelligent, i.e. it possesses a past and a future.

Keywords

MIDI, Digital Signal Processing (DSP), 3D sound, soundmorphing, soundmapping, Fourier Transformation, filters, virtual reality (VR), multimedia, Solaris (S. Lem), plasma, genetic algorithms, organic changes

Description

Architexture is an interactive visual/audio/spatial installation with realtime computer images and realtime sound processing.

Structure

A variable pneumatic-acoustic screen is located in the middle of the room, i.e. a variable rubber skin is stretched over a loudspeaker matrix which can be controlled via a pneumatic air intake. A figure is located on the floor in front of the projection screen: a bullseye, a circle – the center of the action and interaction.

Interaction

1. The first image and the pneumatic screen. The observer, who steps into the cross of the figure on the floor, is scanned with the aid of an infrared camera and is then a part of the pneumatic sculpture: The data on his or her image and movement coordinates are reproduced on the screen as acoustic and topographic information, i.e. the space in front of the screen will be analyzed according to the following criteria with an infrared camera and a directional microphone:
- the observer's relative spatial changes,
- history of the spatial changes,
- sounds produced by the observer,
- history of the sounds.

Ziel

Durch die architektonische + technologische Anordnung wird der Betrachter mit einer variablen, pneumatisch-akustischen Leinwand konfrontiert, wobei sowohl visuelle und akustische als auch räumliche Wahrnehmung angesprochen werden. Die Leinwand selbst verhält sich, als hätte sie eine Intelligenz, d.h. sie besitzt eine Vergangenheit und eine Zukunft.

Keywords

MIDI, Digital Signal Processing (DSP), 3D-Sound, Soundmorphing, Soundmapping, Fourier Transformation, Filter, Virtual Reality (VR), Multimedia, Solaris (S. Lem), Plasma, Genetische Algorithmen, Organische Veränderungen

Beschreibung

Architexture ist eine interaktive Bild/Klang/Raum-Installation mit Realtime-Computerbildern und Realtime-Soundprocessing.

Aufbau

In der Mitte des Raumes befindet sich eine variable, pneumatisch-akustische Leinwand, d.h. über eine Lautsprechermatrix wird eine variable Gummihaut gespannt, die über pneumatische Luftzufuhr gesteuert werden kann. Vor der Projektionswand befindet sich eine Bodengrafik: eine Zielscheibe, ein Kreis – das Zentrum der Aktion und Interaktion.

Interaktion

1. Das Bild und die pneumatische Leinwand: Der Betrachter, der in das Fadenkreuz der Bodengrafik tritt, wird mit Hilfe einer Infrarotkamera ‚abgescannt' und nun selbst Teil der pneumatischen Skulptur: seine Bilddaten und Bewegungskoordinaten werden als akustische und topografische Informationen auf der Leinwand wiedergegeben – d.h. durch Einsatz einer Infrarotkamera und von Richtmikrofonen wird der Raum vor der Leinwand nach folgenden Kriterien analysiert:
- relative Bewegungsänderungen des Betrachters
- Geschichte (History) der Bewegungsänderungen
- erzeugter Schall des Betrachters
- Geschichte (History) des Schalls
Danach werden diese Informationen von einem Steuer-

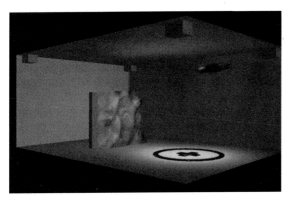

ARCHITEXTURE Raum

computer analysiert und in Impulse umgesetzt, die per Pneumatik die Wand steuern und über die Lautsprechermatrix organisch interpolierten Sound abgeben (history buffered soundmapping). Gleichzeitig wird ein in Echtzeit vom Grafikcomputer generiertes Bild auf die Leinwand projiziert, das wiederum vom Geschehen im Raum abhängt (organisches plasmatisches Spiegelbild). Die Installation ARCHITEXTURE vereinigt also Bild, Bewegung, Ton und Raum zu einer organischen Skulptur.

2. Bewegung, Sprache & Ton: Der Betrachter steht nun vor seinem eigenen, über die Auflösung der variablen Leinwand reproduziertem Abbild. Um sich selber wiederzuerkennen, beginnt er sich zu bewegen, er reckt den Arm, er dreht sich, beugt sich – das Bild in der Projektion reagiert entsprechend. Zur Körpersprache kommt hier jedoch hörbare Sprache hinzu: Ein Mikrophon registriert alle im Raum gesprochenen Worte und Geräusche, sie werden aufgelöst in Klangeinheiten und gespeichert. Die gespeicherten Satz- und Wortfragmente werden vom Computer zu einer künstlichen Sprachmelodie interpoliert. Diese simulierte Sprache wird über ein 3D Soundsystem wiedergegeben. Schließlich erwacht das künstliche Wesen der pneumatischen Skulptur aus seiner nachahmenden Passivität und folgt nicht mehr nur seinem Vorbild. Fast so, als hätten sich die Dimensionen verschoben, als würde der Rezipient zum Spiegelbild, bewegt sich das pneumatische Spiegelbild selbstständig, ‚spricht' zum Betrachter, fordert auf nachzuahmen. Plötzlich reagiert die Installation nicht mehr bloß auf den Rezipienten, der Vorgang kehrt sich um, die Installation ‚spielt' mit ihm, macht ihn zu ihresgleichen, macht ihn zum variablen Spiegelbild.

Invention

Die (softwaretechnische) Implementation einer gehirnähnlichen Struktur in der Zeitachse wird die Interaktion zwischen Betrachter und Computer von ihrer üblichen 1:1 Umsetzung befreien. Das Programm legt also eine Art History an, um Rückschlüsse aus der Vergangenheit des Systems für das zukünftige Verhalten des Systems zu zie-

Afterwards, this information will be analyzed by a control computer and converted into impulses which control the wall via the pneumatic system and emit organically interpolated sound through the loudspeaker matrix (history-buffered soundmapping). At the same time, an image generated by the computer in realtime will be projected onto the screen, and this image will depend on the events in the room (organic, plasmatic reflection). The installation ARCHITEXTURE therefore unites image, movement, sound and space to create an organic sculpture.

2. Movement, language & sound: The observer is now standing in front of his or her own image, which was reproduced through the resolution of the variable screen. In order to recognize him or herself, he or she stretches an arm, turns around, bends over – the projected image will react accordingly. Furthermore, audible language will be added to the body language. A microphone registers all spoken words and sounds in the room; they are broken up into sound units and stored. The stored sentence and word fragments are interpolated by the computer into an artificial language melody. This simulated language will be reproduced via a 3D sound system.

Finally, the artificial being of the pneumatic sculpture will awaken from its imitative passivity and stop following the example of its model. Almost as if the dimensions had shifted, the recipient becomes a reflection; the pneumatic reflection moves independently, "speaks" to the observer, requests him or her to repeat. Suddenly, the installation does not react solely to the recipient; it turns around and "plays" with him or her, makes the observer its equal, a variable reflection.

Invention

The (software) implementation of a structure in the temporal axis which is similar to the brain will free the interaction between observer and computer from its usual 1:1 ratio. The program creates a kind of history in order to obtain conclusions from the system's past for its future behavior (extrapolation). This brain itself has been set up to be "destructive" in its way, i.e. the computer is able to forget and replace superfluous information with meaningful information.

Software

All functions are designed to produce organic behavior which occurs between audio data, digital images and three-dimensional coordinates. In doing so, all modules are linked, i.e. functions can, for example, process audio samples and 3D objects simultaneously. The two-dimensional video originals are transformed into three-dimensional space according to the rules defined by the software. A space-time dimension, a topography of the image so to speak, is created according to this principle: The sequence of similar images in time is no longer of importance, but the number of potential metamorphoses in the space-time component inside the image is. This enrichment of the images by a further dimension, which is linked to an arbitrary dilation of time, makes it possible to free the actual image data from their static state and transfer them to states of greater or less complexity. The additional inheritance of external information (e.g. from the audio range into the visual range) allows the control of the flow of information through parameters which are foreign to the image.

Software methods

Transition from order to chaos / movement:

hen (Extrapolation). Dieses Gehirn selbst ist in seiner Art ‚destruktiv' angelegt, d.h. der Computer ist in der Lage, zu vergessen und überflüssige Informationen durch sinnvolle zu ersetzen.

Software

Sämtliche Funktionen sind darauf ausgelegt, organisches Verhalten zwischen Audiodaten, digitalen Bildern und dreidimensionalen Koordinaten zu erzeugen. Hierbei können sämtliche Module miteinander verknüpft werden, d.h Funktionen können z.B. gleichzeitig Audiosamples und 3D-Objekte verarbeiten. Die zweidimensionalen Video-Vorlagen werden durch software-definierte Regeln in den dreidimensionalen Raum transformiert. Durch dieses Prinzip wird eine zusätzliche Raum-Zeit-Dimension eingeführt, sozusagen eine Topografie des Bildes. Nicht mehr die Abfolge ähnlicher Bilder in der Zeit ist hier von Bedeutung, sondern die Menge der möglichen Metamorphosen in dieser bildinternen Raum-Zeit-Komponente. Diese Anreicherung der jeweiligen Bilder mit einer weiteren Dimension, verbunden mit einer beliebigen Zeitdehnung, ermöglicht es, die eigentliche Bildinformation von ihrem statischen Zustand zu befreien und in Zustände größerer oder kleinerer Komplexität überzuführen. Die zusätzliche Vererbung von externer Information (z.B. aus dem Audiobereich in den Bildbereich) erlaubt es, den Informationsfluß durch bildfremde Parameter zu steuern.

Methoden der Software

Übergang von Ordnung zu Chaos / Bewegung:
– Gravitation, molekulare Dynamik
– Random Walk, Drunken Fly, Würmer...
– chemische Diffusionsvorgänge, Anlagerungsprozesse
– Life Algorithmen
– Überführung von 2D nach 3D (DwarfMorph)
– Soundspezifische Parameter
– Veränderung von Bildelementen im Farbraum

Erzeugung von SubPatterns / SubStrukturen:
– Partikularisierung von Bildern in 2D Objekte
– Partikularisierung von Bildern in 3D Objekte
– Umwandlung von Bildern in Grids (Texturemapping)
– Veränderung von Material- und Textureparametern
– Painteffekte
– 2D/3D Warping (Verzerrung)

Anwendung von Digital Signal Processing (DSP)
– auf 3D Körper, auf 2D Bilder
– auf Sound
– auf Raum-Koordinaten / Bewegungen in der Zeit

Soundspezifische Steuerungen:
– Steuerung von molekularen Bewegungen durch Sound
– Mapping von Audiosignalen auf 3D-Objekte
– Generierung von Formen und Verhaltensmustern,

abhängig von Frequenzen, Amplituden...
– Soundgesteuertes Digital Image Processing

Soundspezifische Parameter:
– Amplitude
– Frequenz / Frequenzbänder – Hüllkurven als Timelines
– Abweichungen, Durchschnitt, Minimum, Maximum,...

Soundgeneration
– Übertragung von 3D-Koordinaten auf MIDI-Parameter (Tonhöhe, Lautstärke,...)
– Übertragung von 3D-Koordinaten auf 3D Lautsprecher-Matrix
– Umrechnung von 3D-Koordinaten in Hüllkurven/Envelopes
– Skalierung von Bewegungspfaden

Metamorphose / Interpolation
– von dreidimensionalen Formen
– von Klängen (Soundmorph)
– von zweidimensionalen Bildern (Morph)
– durch fraktale Algorithmen
– durch Gravitation, Magnetismus,...

Supreme Particles

Anna Bickler	– Frankfurt, Institut für Neue Medien (Video)
Stefan Karp	– Frankfurt (Architektur)
Gideon May	– Amsterdam (Software)
Paul Modler	– Berlin (DSP-Programming, Sounds)
Michael Saup	– Frankfurt, Institut für Neue Medien (Software, Audio)
Rolf van Widenfekt	– Mountainview (Software, Consultant)

SPONSOREN

ArSciMed, Paris
boso, Fabrik medizinischer Apparate, Jungingen
Silicon Graphics GmbH
Städelschule – Institut für Neue Medien, Frankfurt
Steinberg Research, Hamburg
X94, Akademie der Künste, Berlin

– gravitation, molecular dynamics
– Random Walk, Drunken Fly, Worms
– chemical diffusion processes, attachment processes
– life algorithms
– translation from 2D to 3D (DwarfMorph)
– sound-specific parameters
– change of image elements in range of colors

Creation of sub-patterns / substructures:
– particularization of images into 2D objects
– particularization of images into 3D objects
– conversion of images into grids (texturemapping)
– change of material and texture parameters.
– paint effects
– 2D/3D warping (distortion)

Application of Digital Signal Processing (DSP)
– to 3D bodies, to 2D images
– to sound
– to spatial coordinates / movements in time

Sound-specific controls:
– control of molecular movements through sound
– mapping of audio signals onto 3D objects
– generation of forms and behavioral patterns, independent of frequencies, amplitudes...
– sound-controlled Digital Image Processing

Sound-specific parameters
– amplitude
– frequency / frequency bands – generating curves as timelines
– deviations, average, minimum, maximum...

Sound generation
– transfer of 3D coordinates to MIDI parameters (tone pitch, volume...)
– transfer of 3D coordinates to 3D loudspeaker matrix
– conversion of 3D coordinates to generating curves / envelopes
– scaling of movement paths

Metamorphosis / interpolation
– of three-dimensional forms
– of sounds (soundmorph)
– of two-dimensional images (morph)
– through fractal algorithms
– through gravitation, magnetism...

SPONSORS

ArSciMed, Paris
boso, manufacturer of medical apparatuses, Jungingen
Silicon Graphics GmbH
Städelschule – Institut für Neue Medien, Frankfurt
Steinberg Research, Hamburg
X94, Akademie der Künste, Berlin

E-H

MONIQUE MULDER / DIRK LÜSEBRINK / GIDEON MAY

1.
Tolerance of perception in perpectives:

We will see if we can handle various vanishing points in perspectives or, if the complexity increases, we will lose control in chaos. The routes in our thinking concept are always creating several possibilities by one material fact. The perception we have of a point in space and time will undergo a deformation because of our movement in space and time. This will lead us to an unexpected new beginning point.

The best understandable example is that you can see several different perspectives from one viewing point. In principle this implies the openess mankind and nature is capable of. The degree of tolerance in which we handle our surroundings is in direct relation to our behaviour in space and time. This gives us an unlimited reality in our perception and this appears also in our surroundings. Limits are created by focussing. By focussing, the perspectives that are not chosen will deform or will undergo a metamorphosis in the direction of the chosen focus. So it looks like there is a coherent connection between the unconnected diversities. The consequence of our choice will give the possibility of the existence of another world at the same moment in time, which we are not able to perceive. Out of this follows that we constantly have to adjust our perception-tolerance.

The tolerance of our perception that we give to the perspective of our choice should be adjusted during our transformation in time

1.
Die Toleranz in der Wahrnehmung von Perspektiven:

Wir werden sehen, ob wir mit Perspektiven mit mehreren Fluchtpunkten umgehen können, oder ob wir, wenn es sehr komplex wird, im Chaos die Kontrolle verlieren. In unserem Denken werden durch eine materielle Tatsache immer mehrere Möglichkeiten geschaffen. Unsere Wahrnehmung eines räumlichen und zeitlichen Punktes ändert sich, weil wir uns im Raum und in der Zeit bewegen. Das führt uns zu einem unerwarteten neuen Anfang.

Das beste Beispiel ist vielleicht, daß man von einem Punkt aus verschiedene Perspektiven wahrnehmen kann. Im Grunde zeigt das die Offenheit, deren der Mensch und die Natur fähig sind. Das Maß an Toleranz, mit dem wir mit unserer Umgebung umgehen, steht unmittelbar mit unserem Verhalten im Raum und in der Zeit in Relation. Das verleiht unserer Wahrnehmung grenzenlose Realität und taucht auch in unserer Umgebung auf.

Grenzen entstehen durch Fokussierung. Durch die Fokussierung werden die Perspektiven, die nicht gewählt werden, deformiert oder machen eine Veränderung durch in Richtung des gewählten Fokus. Es sieht also so aus, als bestehe zwischen den unzusammenhängenden Diversitäten ein Zusammenhang. Unsere Wahl ermöglicht die Existenz einer anderen Welt in demselben Augenblick, die wir aber nicht wahrnehmen können. Daraus folgt, daß wir unsere Wahrnehmungstoleranz ständig anpassen müssen.

Die Toleranz in der Wahrnehmung, die wir der von uns gewählten Perspektive angedeihen lassen, muß während unserer räumlichen und zeitlichen Transformation angepaßt werden, weil der Blick, der im Moment der Ankunft zu erwarten ist, durch die zeitliche Veränderung aus einer anderen Perspektive kommen wird. Wir müssen also unsere Wahrnehmung anpassen, um dem Unerwarteten Platz zu machen.

Unser Denken gibt uns die Möglichkeit, im Detail zu ersticken, aber auch, die Bedeutung des Details zu erkennen. Beides bringt uns von der Quintessenz weg.

and space, because the expected view at the moment of arrival, by the change of time, will result in another perspective view. The elements force you to adjust your perception to give way to the unexpected.

Our thoughts give us the possibility to get choked in the detail and also to discover the magnitude of the detail. This drives us away from the quintessence.

2.

The sketched installation is setting the viewers in a feedback loop with the given database of the perceptive spaces. The interaction loop starts with the grabbing of videoframes of the exhibition room. In the videoframes the viewer with the candle is detected. That position information is then combined with the applications knowledge about knowledge of the current position of the virtual camera in the perspective space to calculate

the new virtual camera position. By that, the translation of the viewer activities becomes a nonlinear one. One meter in real space can be translated to one milimeter or 10 meters in the perspective space depending on the actual virtual position in the perspective space. Furthermore, the geometry of the database is going to change depending on the viewers activities. New rooms will open up from "D pictures to 3D spaces and others will disappear. This goes beyond a straghtforward navigation in a 3D space, because the space is changing and the mapping from real to virtual space is nonlinear.

Thanks for support:
– ART+COM, Berlin, D
– MATTMO, Den Haag, NL
– Philips Interactive Media, D, European Labels
Manager: Mike Freni
– SOFTIMAGE, Marc Petit

2.
Die Installation versetzt die Betrachter in eine Feedback-Schleife mit der bestehenden Datenbank der Perspektiven-Räume. Die Interaktionsschleife beginnt mit Videorahmen im Ausstellungsraum. In den Videorahmen ist der Betrachter mit der Kerze zu sehen. Diese Positionsinformation wird dann mit dem Wissen um die momentane Position der virtuellen Kamera im Perspektiven-Raum kombiniert, und daraus wird die neue Position der virtuellen Kamera errechnet. Von da an ist die Umsetzung der Handlungen des Besuchers nicht-linear. Ein Meter im realen Raum kann im Perspektiven-Raum als ein Millimeter oder als zehn Meter umgesetzt werden, je nachdem, wo im Perspektiven-Raum sich die virtuelle Kamera dann befindet. Darüber hinaus ändert sich die Geometrie der Datenbank mit den Handlungen des Besuchers. Neue dreidimensionale Räume entstehen aus zweidimensionalen Bildern, andere verschwinden. Das ist mehr als eine Bewegung in einem dreidimensionalen Raum, weil sich der Raum verändert und die Übertragung des realen auf den virtuellen Raum nicht linear ist.

CHAOS CUBE –
INTERAKTIVE MODELLWELT
chaos cube – interactive model worlds

MICHAEL KLEIN

Historische Verknüpfungen

Unsere philosophische und wissenschaftliche Auseinandersetzung mit Modellwelten ist so alt wie das Denken des Menschen in Weltbildern. Der Unverständlichkeit des Kosmos und der Unbegreiflichkeit der komplexen Natur begegneten unsere Vorfahren zunächst mit diversen Götterwelten. Für die westlichen Kulturkreise vermehrten die Griechen die Schöpfungsmythologien um naturphilosophische Erfahrungen. So entstehen die ersten Götter der griechischen Mythologie, Gaia und Eros, nach Anaxagoras [1] aus dem CAOS – dem Tohuwabohu der strukturlosen Mischung – angeregt durch den NOUS – den Geist als einziger nicht mischbarer Entität – mittels „Perichoresis" – einem strukturbildenden Entmischungsprozeß. Platons Höhlengleichnis, welches ein Verständnis der Natur als Problem der Interpretationen von Projektionen der Realität durch den menschlichen Beobachter formuliert [2], markiert bis heute unser erkenntnistheoretisches Dilemma. Unsere modernen naturwissenschaftlichen Weltbilder entstanden aus der Auseinandersetzung mit dem Uhrwerk-Universum des mechanistischen Zeitalters. Haben uns die modernen physikalischen Theorien – Relativitätstheorie und Quantenmechanik – auch von der Vorstellung der absoluten Raum-Zeit sowie der Beobachterunabhängigkeit des physikalischen Kosmos befreit, so glauben wir doch mit Pythagoras und Galilei, daß „die Zahl die Natur und das Wesen der Dinge ist". Die Sprache zum Verständnis der Natur ist die mathematisch formalisierte Physik [3]. In diesem Sinne ist die Natur deterministisch berechenbar und formal mathematisch begreifbar. Unsere aktuellen Weltbilder sind geprägt von den virtuellen Realitäten des Computerzeitalters [4]. Finden sich die frühesten und originellsten Beschreibungen der Idee virtueller Welten auch in Science-Fiction-Romanen, wie „Simulacron III" [5,6], so ist das Konzept den vermeintlich objektiven Naturwissenschaften nicht fremd. Im Gegenteil, ist doch die Grundlage der klassischen Naturwissenschaften die Idealisierung natürlicher Prozesse durch simulierte Modellwelten. Einstein benutzte für ihre Formulierung Gedankenexperimente, heute

Historical Connections

Our philosophical and scientific exposition with model worlds seems to be as old as mankind thinking about views of life. Our ancestors dealt with the unintelligibility of the cosmos and the unconceivability of nature by inventing a variety of deities. For the western cultures the Greeks added philosophical experiences about nature to the mythologies of creation. According to Anaxagoras [1] the first deities of the Greek mythology Gaia and Eros arose from CAOS -the tohuwabohu or chaos without any structure- initiated by the NOUS -the spirit understood as the only non-miscible entity- by a process he called "perichoresis" -developing structures by a reverse mixing operation-. Plato's image of the cave, where he relates the problem of understanding nature to the problem of a human observer interpreting projections of reality [2], still documents our dilemma with any theory of cognition. Our modern scientific views of the cosmos are founded on the discussions of the mechanical epoch about the image of the clockwork universe. Though the modern physical theories -like the theory of relativity and quantum mechanics- have liberated us from the idea of absolute space-time and the observer independency of the physical universe, we still believe with Pythagoras and Galilei that "the number is the character and essence of reality". The language for understanding nature is the mathematically formalized theory of physics [3]. In that sense nature may be deterministically calculable and mathematically understandable. Our actual understanding of reality is significantly influenced by the virtual realities of the computer age [4]. Though the first and most original descriptions of virtual worlds can be found in science fiction literature, such as "Simulacron

III"[5, 6], the concept is very well-known in the natural sciences. The foundation of classical sciences is the idealization of natural processes with simulated model worlds. Einstein handled them as "Gedanken"-experiments; today our model worlds can be numerically simulated with computers. The Chaos Cube is a computer-graphical interactive installation. It is an experiment for the theory of cognition. Using an example of a pure abstract mathematical model world it illustrates the problems of dealing with "intelligent ambiences". The installation does not concentrate on real time photorealism for I believe the main challange is the understanding of novel scenarios.

Mathematical Model World

The Chaos Cube allows for an interactive i.e. visual immersion in the virtual world of the socalled Chaotic Hierarchies [7]. These are time-continuouse and time-discrete mathematical equations which simulate all kinds of nonlinear dynamics. The chaotic hierarchy uses the concept of hierarchical mathematical systems. The Chaos Cube can handle from two to four embedding dimensions. For observation the dynamical states, which are the solutions of the mathematical equations depending on the initial conditions, are projected as sets of points or trajectories of the attractors. Different dynamical states correlate with significant geometrical structures. Simple periodical and therefore regular dynamics will give finite sets of points or closed orbits. Chaotic states show complex clouds of points or non-repetitive line structures.

lassen sich unsere Modellwelten vollständig algorithmisch in Computern simulieren. Der Chaos Cube ist als computergraphische interaktive Installation ein erkenntnistheoretischer Versuch, die Probleme im Verständnis „intelligenter Ambiente" am Beispiel rein abstrakter, mathematischer Modellwelten aufzuzeigen. Bewußt sucht diese Installation keine Nähe zum Echtzeit-Photorealismus, denn ich glaube, die eigentliche Herausforderung liegt im Begreifen neuartiger Szenarien.

Mathematische Modellwelt

Der Chaos Cube erlaubt ein interaktives, visuelles Eintauchen in die virtuelle Welt der sogenannten Chaotischen Hierarchien [7]. Dies sind zeit-kontinuierliche und -diskrete mathematische Gleichungen, die alle Typen von nichtlinearer Dynamik simulieren können. Die chaotische Hierarchie benutzt die Idee hierarchischer mathematischer Systeme, der Chaos Cube kann z.B. zwei bis vier räumliche Einbettungsdimensionen darstellen. Für den Beobachter

werden die dynamischen Zustände, welche den Lösungen der mathematischen Gleichungen, abhängig von den jeweiligen Anfangsbedingungen, entsprechen, als Punktmengen- bzw. Linien-Attraktoren im mathematischen Zustandsraum projeziert. Qualitativ verschiedene dynamische Zustände unterscheiden sich durch jeweils verschiedene geometrische Strukturen. Einfache periodische, regelmäßige Dynamiken stellen sich als endliche Punktmengen oder geschlossene Orbits dar. Chaotische Dynamiken zeigen komplexe Wolken von Punktmengen oder komplizierte, sich nie wiederholende Linienstrukturen.

Interaktive Manipulationen

Auf der obersten Ebene wählt der User das gewünschte Modellsystem als zeitdiskrete Iterationsgleichung oder zeitkontinuierliche Differentialgleichung sowie die Dimension des entsprechenden Modells zwischen zwei und vier Freiheitsgraden. Danach wandert er mit einem Cursor über eine zweidimensionale Parameterebene und selektiert den gewünschten Parametersatz. Die Colorierung der Parameterebene und die Attraktor-Projektionen auf den Seitenwänden des Chaos Cube erlauben eine gezielte Differenzierung. Die dynamische Lösung zum selektierten Zustand findet der User als stereographische Projektion vor sich schwebend. Diese globalen Ansichten der Attraktoren kann er frei im Raum rotieren oder sich in Ausschnitten herauszoomen lassen. Die Idee der Endo-Space Science [1] postuliert die faszinierende Möglichkeit, durch Veränderung unserer Beobachterposition zum Objekt andere und eventuell neue Erfahrungen über unsere Welt zu gewinnen. Der Chaos Cube ermöglicht es, die lokale Objektwelt

Interactive Manipulations

On the highest level of interaction the user chooses the model system from time-discrete iterative maps or time-continuous differential equations and chooses the dimension of the model between two to four degrees of freedom. With the help of a cursor he walks around a two-dimensional parameter plane and selects the parameter values. The coloring of the parameter plane and the projections of the attractors on the sidewalls of the Chaos Cube helps with the differentiation. The dynamical solutions according to the choosen state can be seen as floating stereographical projections. These global views of the attractors may be freely rotated and zoomed in. The idea of Endo-space science [1] postulates the fascinating opportunity to change our position of observation to the objects of interest to probably gain new insight of our world. With the Chaos Cube it becomes possible to view the local object world from an insite view. Dynamical systems may be visualized from the local space of Eigenvectors. The Chaos Cube can simulate a point of observation right on the trajectories of the system, i.e right on the flow evolving in time. The observer finds himself within a local volume of space which follows the trajectory. It is continuously deformed by the forces acting upon the local environment. The observer experiences the time-dependent and structural evolution of the system on site.

Manipulator Interface

The interactive control of the Chaos Cube is realized with a novel manipulator. The interface is a crotch the user carries around. The manipulator is fitted with a small keyboard and a microphone for controling the system with a simple command language. The position of the manipulator in the space in front of the projection screen is tracked with an ultrasonic device. This allows for a location of the position on the virtual parameter plane and an observer centered geometrical projection of the objects. The active system parameters, some control commands and online help for system control, will be displayed on an additional screen.

von einem inneren Beobachtungsstandpunkt aus zu betrachten. Dynamische Systeme lassen sich im lokalen Eigenvektorraum visualisieren. Der Chaos Cube bietet die Möglichkeit, die Beobachterposition direkt auf die Trajektorie des Systems, auf den sich in der Zeit verändernden Fluß zu verlegen. Der Beobachter befindet sich im Inneren eines lokalen Raumvolumens, das sich auf den Trajektorien mitbewegt und sich entsprechend der Kräfte, die in der lokalen Umgebung herrschen, kontinuierlich deformiert. Der Beobachter erlebt „vor Ort" die zeitliche und strukturelle Evolution des Systems.

Manipulator Interface

Die interaktive Kontrolle über den Chaos Cube geschieht mit Hilfe eines neuartigen Manipulatorstabes. Der User hat als Interface diesen Manipulatorstab in der Hand, an dem sich ein Mikrofon für die Systemkontrolle mittels einer einfachen Kommandosprache befindet sowie ein kleines Tastaturfeld mit entsprechenden Steuerungstasten. Die Position des Manipulatorstabes im Raum vor der Projektionsfläche wird mittels eines Ultraschallsystems ausgewertet und erlaubt so einerseits die Parameterauswahl in der virtuellen Parameterebene wie auch eine beobachterzentrierte geometrische Darstellung der Objekte. Über einen seitlich aufgestellten Kontrollschirm werden die gewählten Systemparameter und die Kontrollkommandos dargestellt sowie Hilfestellungen zur Systemkontrolle angeboten.

LITERATURE

[1] O.E. Rössler, Endophysik, P. Weibel Hrsg., Merve Verlag, Berlin 1992
[2] Platon, Politeia, Ges. Werke III, 7. Buch
[3] K. Simonyi, Kulturgeschichte der Physik, Urania Verlag Leipzig 1990
[4] Kataloge Ars Electronica 1990-1994
[5] F.F. Galouye, Simulacron III, Heine Verlag 1967
[6] R.W. Faßbender, Die Welt am Draht, Film 1973
[7] G. Baier, M. Klein, A Chaotic Hierarchy World Scientific Singapore 1991

LITERATUR

(1) O.E. Rössler, Endophysik, P. Weibel Hrsg., Merve Verlag, Berlin 1992
(2) Platon, Politeia, Ges. Werke III, 7. Buch
(3) K. Simonyi, Kulturgeschichte der Physik, Urania Verlag Leipzig 1990
(4) Kataloge Ars Electronica 1990-1994
(5) F.F. Galouye, Simulacron III, Heine Verlag 1967
(6) R.W. Faßbinder, Die Welt am Draht, Film 1973
(7) G. Baier, M. Klein, A Chaotic Hierarchy, World Scientific Singapore 1991

ELECTRONIC MIRROR 2

CHRISTIAN MÖLLER

Spiegel verleihen Gewißheit – eine Banalität, die sich spätestens bei der Morgentoilette tagtäglich aufs neue bewahrheitet. Von Electronic Mirror 2 wird diese Banalität auf den ersten Blick widerlegt. Ähnlich den Zerrspiegeln der Jahrmärkte inszeniert diese Spiegelinstallation nämlich weniger das genaue Konterfei des Betrachters als seinen bewegten, irritierten Kampf ums eigene Bild. Doch im Unterschied zum Szenario der Jahrmärkte vollführt der Betrachter in Electronic Mirror 2 seine Experimente in einem markierten Feld: in einem Spiegel-Stadion, das der Betrachter durchlaufen muß, um ein scharfes Bild seiner selbst zu erlangen.

Abgesteckt ist das Spiegel-Stadion auf der einen Seite durch einen Spiegel, dessen Transparenz und Schärfe variabel ist, und auf der gegenüberliegenden Seite durch eine Ebene, in die die Laserdisksequenz eines betrachtenden Auges eingespiegelt ist. Zwischen Spiegel und Auge bewegt sich der Betrachter, der nach einem getreuen Abbild seiner selbst unterwegs ist. Wenn er sich dem Goldrandspiegel nähert, um sich selbst schärfer zu fassen, wird er feststellen, daß sein Spiegelbild verschwimmt. Verantwortlich für diese diffuse Selbsterfahrung ist die Vernetzung eines Ultraschall-Sensors, der die Entfernung des Betrachters zum Spiegel mißt, mit einem PC, der über einen Dimmer die Transparenz und Abbildungsschärfe des Spiegels reguliert. Unmittelbar vor dem Spiegel kriegt der Betrachter sein Bild nicht zu fassen. Der Gegenstand ist – mit Kierkegaards Wort – der Sehnsucht zu nahe, „so nahe, daß er in ihr ist".

Es ist auch die Erfahrung des mythologischen Spiegelklassikers, die diese variable Fokussierung des Selbstbilds inszeniert: Auch für Narkissos war das eigene Spiegelbild eine überraschende und späte Erfahrung seines Wanderns. Eine Erfahrung, die ihm wie dem Betrachter von Electronic Mirror 2 unterwegs zuteil wurde. Dieser Urszene einer Spiegelerfahrung und ihrer tödlichen Kosequenz begegnet Electronic Mirror 2 jedoch computergesteuert mit Entzug: Das Spiegel-Stadion verhindert das selbstverliebte Hineinfallen ins eigene Bild. Das Wiedererkennen des Selbst an einem bestimmten Ort der Blickachse setzt Distanz voraus und ist in eine andere Urszene eingelassen: Es ist gekoppelt mit dem zusehenden, anerkennenden Auge eines anderen. Tatsächlich liegt der Punkt, der das Selbstbildnis des Betrachters in voller Schärfe zeigt, genau auf der Mittelachse des Raums, bei dessen Betreten sich das Auge öffnet.

SUSANNE CRAEMER

Mirrors provide certainty – a banality which proves itself anew every day, at the latest during the morning toilet. Electronic Mirror 2 refutes this banality at first glance. In a way similar to the mirrors in funhouses, this installation of mirrors produces, rather than an accurate likeness of the viewer, his or her animated, irritated struggle to find the true image. However, in contrast to funhouses, the viewer's experiments in Electronic Mirror 2 are in a marked field, a mirrored stadium which must be passed in order to obtain a clear image of him or herself.

The mirrored stadium is defined on one side by a mirror, the transparency and clarity of which can be varied, and by a flat surface on the other side, in which the laser-disk sequence of an observing eye is shown. The viewer, searching for an accurate image of him or herself, moves between the mirrors and the eye. When he or she approaches the mirror with the golden frame to focus, the mirrored image will blur. The reason for this vague experience of self is the connection between an ultrasonic sensor which measures the viewer's distance to the mirror and a PC which regulates the mirror's transparency and the reflective clarity via a dimmer. Directly in front of the mirror, the viewer cannot focus his or her image. The object is – in Kierkegaard's words – too close to the longing, "so close that it is within it".

The experience of the classic myth is also the source of this variable focus of the self-image: Even Narcissus' own mirrored image was a surprising experience late in his wanderings. An experience while he was in motion, just as the viewer of Electric Mirror 2. This ur-scene of an experience with a mirror and its deadly consequence is met by Electric Mirror 2 with deprivation under computer control: The mirrored stadium hinders the viewer from falling into the self-image which he or she is in love with. The recognition of the self at a certain place in the gaze's axis requires distance and is set in another ur-scene: It is coupled to the observing, acknowledging eye of another. In fact, it is exactly the point on the room's central axis where one must stand in order to open the eye, which shows the viewer's self-image in all its clarity.

LC-Spiegel

Transparent Opak

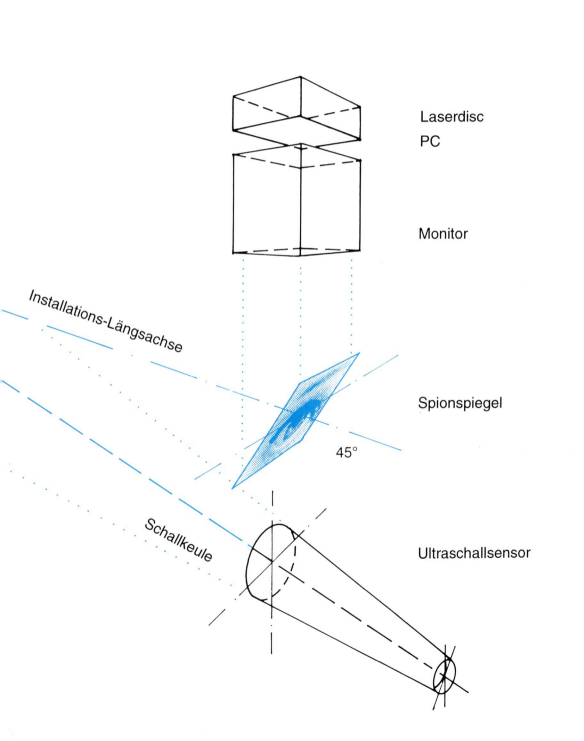

AS FAR AS IN-BETWEEN

MARTIN KUSCH

Absence and Transparency

One credo of European rationality is transparency. Its claim to analysis, understanding and construability of the world is formulated in this way. Reason may make the motives and purposes of all actions, facts and events obvious, transparent. In the imagery of light, this metaphysical system of illumination, this claim to enlightenment, is expressed visually. Consequently, "let there be light" means virtually the same as "let there be truth". In the glade of being, the hidden truth will become evident. "Let there be light", however, also means virtually the same as "the world shall be created". Reflections of light and the play of light are therefore reproduction media only in a limited, formal interpretation of the term. In its true meaning, art consisting of light actually denotes constructive creation. Therefore, the artist is, however, not a creator mundi him or herself alone, but the one who sees the light, the beholder, also plays this role. With a photograph, the beholder becomes a cocreator, an equal partner in the act of creation. There are various manifestations of light: the various media in which it shows itself. From a spotlight to a film projector, from the camera to the screen, we can find sources of light and surfaces of light which have various spatial and temporal functions. Light produces not only one world, but multiple worlds. Color is one world which light gives us; a second is that of movement. The illusory world of light, the theater of movement in the tent of time which we term the art of moving images, is borne of these worlds. A dynamic system between the movement of the body and the movement of the light is created. From the spotlight to the screen and the monitor. When regarding an image, the shadow of the beholder's body produces the image itself. However, it creates not only an image in the world of the here and now; on the contrary, the be-

Absenz und Transparenz

Ein Credo der europäischen Rationalität ist die Transparenz. Damit wird ihr Anspruch auf Analyse, Verstehen und Konstruierbarkeit der Welt formuliert. Die Ratio möge die Motive und Zwecke aller Handlungen, Fakten und Ereignisse durchschaubar machen. In der Lichtmetaphorik kommt diese Metaphysik der Erhellung, dieser Anspruch der Aufklärung visuell zum Ausdruck. „Es werde Licht" heißt demnach soviel wie „es werde Wahrheit". In der Lichtung des Seins werde die verborgene Wahrheit offenkundig. „Es werde Licht" heißt aber auch soviel wie „die Welt sei erschaffen". Lichtreflexionen und Lichtspiele sind also nur in einer eingeschränkten formalen Interpretation Medien der Reproduktion. Lichtkunst bedeutet vielmehr im eigentlichen Sinne konstruktive Kreation. So ein Creator Mundi ist aber nicht allein der Künstler selbst, sondern auch derjenige, der das Licht sieht, also der Betrachter. Der Betrachter wird beim Lichtbild zum Mitautor, zum gleichberechtigten Partner des Schöpfungsaktes. Nun gibt es verschiedene Erscheinungsformen des Lichtes, die verschiedenen Medien, in denen es sich zeigt. Vom Scheinwerfer zum Videoprojektor, von der Kamera zum Monitor finden wir Lichtquellen und Lichtflächen vor, die verschiedene Raum- und Zeitfunktionen haben. Das Licht entwirft nicht nur eine Welt, sondern multiple Welten. Die

Farbe ist *eine* Welt, die uns das Licht schenkt, eine zweite ist die der Bewegung. Aus diesen entsteht die Scheinwelt des Lichts, das Theater der Bewegung im Zelt der Zeit, das wir die Kunst des bewegten Bildes nennen. Es entsteht ein dynamisches System zwischen der Bewegung des Körpers und der Bewegung des Lichts. Vom Scheinwerfer über die Leinwand zum Bildschirm. Der Schatten des Körpers des Betrachters beim Betrachten eines Bildes erzeugt das Bild selbst. Er erzeugt aber nicht nur ein Bild in der Welt des Hier und Jetzt, sondern durch die Scheinwelten der technischen Medien im Theater des Schattens und des Lichts erzeugt der Körper des Betrachters viele Bilder in vernetzten multiplen Bildwelten. Ein Kalkül errechenbarer Ereignisse mutiert zu einem Kalkül vernetzter Bilder in mehrfach gebrochenen Ereignisräumen von verschiedener optischer und ontischer Dichte. Was Martin Kusch durch seine dislozierte Installation in die Kunst einführt, ist die Unterscheidung von optischer und ontischer Dichte. Da wir im natürlichen Raum gewohnt sind, in der Einheit eines Raumes und eines Lichtkontinuums zu leben, haben wir, wie anfangs skizziert, eine Metaphysik des Lichts entworfen, wo ein Ort eine Wahrheit zeigt, also eine Identität von Optik und Ontologie. Kusch zerbricht nun diese Ontologie; verschiedene Lichtquellen an verschiedenen Orten erzeugen verschiedene Bilder der Wahrheit und der Welt. Vor allem ist es der Körper des Betrachters, der als Drehscheibe zwischen diesen multiplen Bildwelten fungiert, diese verschiedenen Bildwelten konstruiert und steuert. Seine Installation zeigt uns nicht nur, wie aus dem Nichts, aus dem bloßen Beobachtungsakt ein Kunstwerk entsteht, sobald es in ein entsprechendes Beobachtungsenviron-

holder's body produces many images in the networked multiple worlds of images, and this is made possible by the technical media's illusory worlds in the theater of shadows and light. A calculation of computable events mutates to a calculation of networked images in multiply refracted spaces of various optical and ontic densities. With his dislocated installation, Martin Kusch introduced the differentiation between optical and ontic density to art. Since we are accustomed to living in natural space in the unity of a space and of a light continuum, we have designed a metaphysical system of light as described above where a place shows a truth, in other words an identity of optics and ontology. Kusch destroys this ontology, various light sources at various locations create various images of truth and the world. Above all, it is the body of the beholder which functions as a switch between these multiple worlds of images, which constructs and controls these worlds. His installation shows us not only how, as if from nowhere, from the simple act of looking, a work of art is created as soon as it has been embedded in an appropriate environment, in an appropriate measuring chain forming a network in itself; even more, however, it shows us an example of contextual art, i.e. how the context, which consists of variables controlled by the act of looking, creates a text, a product. The installation is a system which is dynamic in many ways, which

AS FAR AS INBETWEEN
Eine interaktive Videoinstallation

Raumsituation:

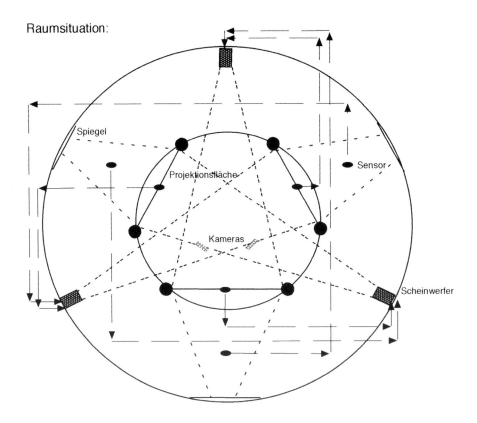

Die Videoprojektionen finden über Spiegel statt
Die beiden Kreise kennzeichnen die Aktionsräume, es gibt in der
Installation zwei Aktionsräume, einen Inneren und einen Äußeren.
Die gestrichelten Linien stellen Projektions- und Lichtkanäle, dar in
denen sich der Betrachter bewegt.
Die 6 Sensoren bilden 2 Gruppen und sind mit den Scheinwerfern vernetzt:
a) innerer Raum - lichtempfindliche Sensoren
b) äußerer Raum - Bewegungsmelder
Die Vernetzung wird durch gestrichelte Linien dargestellt.

Signalverlauf:

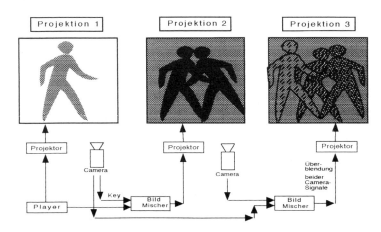

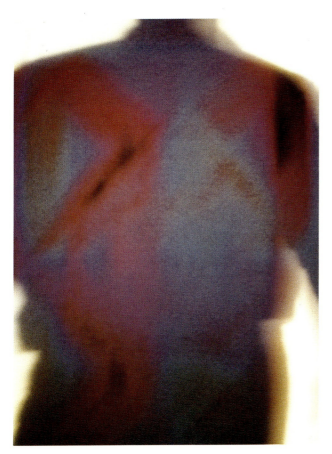

stimulates itself, auto-catalytic, auto-poetic. A refracted symmetry of beholder and image, of body and abstraction, of reality and illusion which creates an optical density without ontology. The images delete and negate themselves, paradoxically creating new images. The zones of the invisible contain single islands of the visible. The images are delocalized and decentralized and become layers of a variable visibility.

PETER WEIBEL

ment, in eine entsprechende, in sich vernetzte Meßkette eingebettet ist, sondern sie zeigt uns vor allem ein Beispiel von Kontextkunst, d.h: wie der Kontext aus beobachtungsgesteuerten Variablen einen Text, ein Produkt erzeugt. Die Installation ist ein mehrfach dynamisches System, das sich selbst erregt, autokatalytisch, autopoetisch. Eine gebrochene Symmetrie aus Beobachter und Bild, aus Körper und Abstraktion, aus Wirklichkeit und Schein, die eine optische Dichte ohne Ontologie erzeugt. Die Bilder löschen und negieren sich selbst, dadurch entstehen paradoxerweise neue Bilder. Die Zonen des Unsichtbaren erhalten partikuläre Inseln des Sichtbaren. Die Bilder werden delokalisiert und entzentralisiert und werden zu Schichten variabler Visibilität.

PETER WEIBEL

Mit freundlicher Unterstützung:

DER INTELLIGENTE BRIEFTRÄGER
the intelligent mailman

MICHAEL BIELICKY

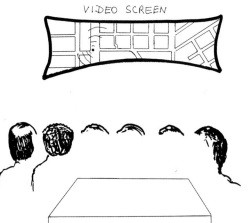

Over many centuries, the mailman (messenger) has become the archetype of an information carrier. His job has remained unchanged to the present day. He still delivers information (letters) by hand. The GPS (Global Positioning System) developed by the American Department of Defense has been extended over the entire globe. Of all the artificial ambiences we have at present, it is the most global. With the aid of the GPS, one can determine his or her location relatively precisely at almost any spot on the Earth in three dimensions (length, width, height). The control system consists of 24 satellites that alternately send their signals to the Earth. When one links the GPS to a sender (e.g. a cellular telephone), one can transfer its data to other places. By giving the information carrier (mailman) a GPS/cellular phone unit and having him run through the urban landscape, one can make his movements visible in the form of a virtual mailman at another location (e.g. a museum). In doing so, this "intelligent mailman" becomes an anachronism: On the one hand, he still delivers the information by hand; on the other hand, he is able to simultaneously transmit "live" information on his location and movements to other localities across any distance. With the aid of such a system, humans could become computer mice. I am interested in developing this system further, thereby organizing virtual performances, for example. I can well imagine that, as a performer, one could control all kinds of devices (robots, virtual spaces, etc.) because of the changes in location (latitudinal and longitudinal coordinates).

"The Intelligent Mailman", in his anachronistic dual function, represents the present intermediate stage on the way to the total age of information.

Der Briefträger (der Bote) wurde über viele Jahrhunderte zum Archetyp eines Informationsträgers. An seiner Tätigkeit hat sich bis heute nichts verändert. Er überbringt noch immer die Informationen (Briefe) per Hand. Das vom amerikanischen Verteidigungsministerium entwickelte Navigationssystem GPS (Global Positioning System) ist über den gesamten Erdball gespannt. Es ist das globalste künstliche Ambiente, das wir bis jetzt zur Verfügung haben. An praktisch jedem Ort der Erde kann man ziemlich genau mit Hilfe des GPS seinen eigenen Standort bestimmen und zwar in dreidimensionaler Form (Länge, Breite, Höhe). Das Steuerungssystem besteht aus 24 Satelliten, die abwechselnd ihre Signale zur Erde senden. Wenn man das GPS-Gerät mit einem Sender (z.B. einem Funktelefon) verbindet, kann man die Daten des GPS zu anderen Orten übertragen. Wenn man dem Informationsträger (Briefträger) solch eine GPS-Funktelefon Einheit in die Hand gibt und ihn durch die urbane Landschaft laufen läßt, kann man an einem anderen Ort (z.B. Museum) seine Bewegung in Form eines virtuellen Briefträgers sichtbar machen. So wird dieser „intelligente Briefträger" zum Anachronismus: denn auf der einen Seite überbringt er die Information noch immer per Hand von Ort zu Ort, aber gleichzeitig ist er imstande, eine Information über seinen Standort und seine Bewegung über jede Entfernung und auch zu mehreren Orten „live" zu übertragen. Mit Hilfe eines solchen Systems könnte der Mensch zu einer Computermaus werden. Mein Interesse ist es, dieses System weiterzuentwickeln und so auch z.B. virtuelle Performances zu veranstalten. Ich kann mir gut vorstellen, daß man als Performer aufgrund der Veränderung seines Standortes (Längen- und Breitenkoordinaten) alle möglichen Vorrichtungen (z.B. Roboter, visuelle Räume, usw.) steuern könnte.

„Der intelligente Briefträger" in seiner anachronistischen Doppelfunktion stellt das heutige Zwischenstadium auf dem Weg zum kompletten Informationszeitalter dar.

MICHAEL BIELICKY

Technical description

A portable GPS device is linked to a cellular telephone containing a modem. Carrying this device, a person (e.g. a mailman) runs through the city.

Technische Beschreibung

Ein tragbares GPS-Gerät wird mit einem Funktelefon – darin ist ein Data-Modem eingebaut – verbunden. Mit dieser Vorrichtung läuft eine Person (z.B. Briefträger) durch die Stadt. Die Daten werden zu einem anderen Ort (z.B. Museum) übertragen. Dort werden diese Daten über die Telefonleitung in einem PC, in welchem eine digitale Stadtkarte des Ortes gespeichert ist, aufgearbeitet. Dann wird auf einer Projektionsleinwand mit Hilfe eines Videobeams die Situation sichtbar gemacht.

The data are transmitted to another location (e.g. a museum) via the telephone line. These data are then processed in a PC in which a digital city map is stored. Lastly, a video beam makes the situation visible on a projection screen.

Besonderer Dank für die freundliche Unterstützung geht an die Firma COMMUNIKATION & NAVIGATION aus Grieskirchen, die die GPS-Hard- und Software zur Verfügung gestellt hat. Das GPS (Global Positioning System), mit welchem sich die Firma beschäftigt, ermöglicht ein zeit-, sicht- und wetterunabhängiges Navigationsverfahren, welches im Bereich Navigation, Vermessung, digitale Landkartentechnologie und Fahrzeugortung Anwendung findet. Herzlichen Dank auch an die Software-Entwicklungsfirma CEO aus Linz, die zusätzliche Hard- und Software für das Projekt zur Verfügung gestellt hat.

Golden Calf

JEFFREY SHAW

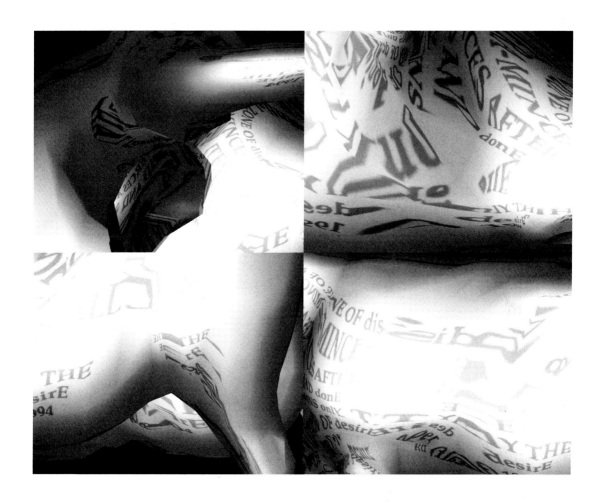

THAT inhabitant ZONE OF disillusion -
THE INTRA-MINCE-
IN THE INTERCES AFTER ALL IS said AND done
THERE REMAINS only THE REFLECTION OF desire
(the Golden Calf) 1994

BROTHERHOOD – TABLE III

Eine Serie interaktiver Konstruktionen
a series of interactive constructions

WOODY VASULKA

Eine Erklärung

Brotherhood ist eine Zusammenfassung verschiedener Medienkonzepte und Ausdruck bestimmter ideologischer und persönlicher Anliegen. Das zentrale Thema von *Brotherhood* kreist zwar um das Dilemma *männlicher* Identität, doch kann man es auch so sehen, daß es aus dem allgemeinen Zwang der Menschheit entstand, die Natur umorganisieren zu wollen. Dieser Prozeß zerstört selbstverständlich die natürliche Ordnung und führt zu Polarisierungen und Antagonismen in verschiedenen gesellschaftlichen und philosophischen Bereichen. Das *Männliche* wird wieder in den Kontext des Krieges als einem zu erwartenden und integralen Bestandteil der menschlichen Evolution gestellt, in den Kontext der Ersinnung und Verwerfung menschlicher Utopien und der gefährlichen Werte männlicher Sexualität. Meine Arbeit tritt nicht für eine reformistische Position oder Verteidigungsstrategie ein. Sie steht verständnisvoll auf der Seite des *Männlichen,* kann einem ironischen Seitenblick auf dessen klar selbstzerstörerisches Schicksal aber nicht widerstehen.

Diese Arbeit vermeidet einzelne Disziplinen, Genres oder Stile und spürt dafür Cluster systemischer Ausdrucksprimitiva auf – solche, die menschen-ähnlich scheinen, aber maschineninhärent sind.

Die Arten unterschiedlicher Medienpartizipation, die in *Brotherhood* vorkommen, lassen sich deshalb kaum beschreiben, weil sich Begriffe wie menschliche oder künstliche „Intelligenz" nicht interpretieren lassen. Während bestimmte elektromechanische Systeme ein *Volumen* an kulturellem Gut enthalten können, kann ihre Brauchbarkeit und ihr Wert nur vor dem Hintergrund der Ausbildung bestimmter menschlicher Verhaltensweisen oder Verfahrensrituale beurteilt werden. Ein Beispiel: *Table IV* kann, gesteuert durch eine menschliche Stimme aus dem Mikrophon, digitale Stimmerkennung und Plotter, tatsächlich physisch einen Brief schreiben, und diese ziemlich komplizierte menschliche Tätigkeit emulieren. In der Vergangenheit verwendete man den Begriff *Intelligenz* einfach für ähnliche Fähigkeiten der Maschinen. Inzwischen ist er aber so bedeutungsgeladen, daß man prosaischer von *einfachen Verhaltensmustern* oder, zur Zeit sehr in Mode, von den *neu entstehenden Eigenschaften komplexer dynamischer Systeme* spricht.

Brotherhood ist aber ein abstraktes Stück und läßt sich nicht korrekt analysieren. Wenn in diesem Kontext die

A Statement

The *Brotherhood* is a summary of media concepts presenting a specific domain of ideological and personal concerns. While the central theme of the *Brotherhood* revolves around the dilemma of *male* identity, it could be understood as arising from the general compulsion of mankind to re-organize Nature itself. This process is of course destructive to the natural order and leads to conditions of polarization and antagonism in various social and philosophical strata. It presents the *male* once again in the context of warfare as an expected and integral part of human evolution, in the construction and abandonment of human utopia, in perilous values of male sexuality. This work does not argue for a reformist agenda or a strategy of defence. It stands sympathetically on the side of the *male* but it cannot resist an ironic glance at his clearly self-destructive destiny.

While avoiding a single discipline, genre or style, the work tends to track clusters of systemic expressive primitives – those which seem human-like yet reside within the machine.

The frustration at describing the modes of various media participation found in *Brotherhood* lies in a generic failure to interpret concepts like human or machine "intelligence." While certain electro-mechanical systems can contain a *volume* of cultural property, their usefulness or value can only be judged against the exclusive domain in modelling of certain human behavior or in acquisition of procedural rituals. For example: *Table IV* possesses the faculty of performing physical letter writing under the control of a human voice via microphone, digital voice recognition and plotter system, thus emulating this rather complex human activity. In the past, the term *intelligence* would have been freely applied to the similar machine state. Burdened with too much meaning, it is being replaced by more somber terminologies such as *simple behavioral patterns* or the more fashionable *emergent properties of complex dynamical systems*.

But the *Brotherhood* is after all an abstract

piece and does not lend itself to correct analyses. If art should participate in this context, the authentic technological extensions and constraints will clearly impose themselves on the work. As of yet this is the most complex work I have attempted with requisite knowledge of various crafts: electronics, optics, engineering and computer programming.

The Tables
(General Description)

Project *Brotherhood* is a complex assembly of six smaller arrangements acting in a mutually coordinated manner as a series of Tables.

The *Tables* are quadratic cage arrangements placed horizontally on metal table legs. Each *Table* contains instruments, able to produce, compose and display varied acoustic and visual structures. Additionally these clusters of technology exhibit a certain volume of behavior through digital programs or in reaction to a set of sensors associated with each *Table*.

Table III
(Functional Description)

Table III holds two picture delivery arrangements: the first is a specialized slide projector while the other is a video projector. Each of these systems is associated with a family of images that occupy a specific projection environment: the stills are confined to a small six-screen layout while the moving images occupy an extended projection environment. Both kinds of projections share the identical pathway of a six-way beam splitter with the images distributed along six axes of cubical vectors to the six screens. During the still image sequence, the projection is intercepted by smaller screen/ frames defining its own projection environment out of the general space. These small frames fold, freeing the projection path for the moving image sequence. This extended projection environment is defined by an arrangement of six projection screens, four standing on the floor plus one suspended from the ceiling. The character of the screen material lets the images appear on both sides, extending the installation's observation mode from the inner core to the outside. There the installation becomes an object with a multitude of interrelated images. The installation has additional functional elements of sound and interactivity. These provide a mode for determining the observer's presence and a certain level of participation.

Kunst eine Rolle spielen sollte, dann werden sich die authentischen technologischen Erweiterungen und Beschränkungen im Werk klar bemerkbar machen. Diese Arbeit ist die komplexeste, die ich bisher gemacht habe, und basiert auf Kenntnissen in der Elektronik, in der Optik, im Maschinenbau und in der Programmierung von Computern.

Die Tische (Allgemeine Beschreibung)

Das Projekt *Brotherhood* ist eine komplexe Anordnung kleinerer Einheiten, die als Serie von Tischen koordiniert agieren.

Diese *Tische* sind quadratische Anordnungen, die horizontal auf metallenen Tischbeinen sitzen. Auf jedem *Tisch* gibt es Instrumente, die verschiedene akustische und visuelle Strukturen produzieren, komponieren und darstellen können. Zusätzlich zeigen diese Technologiecluster durch digitale Programme oder als Reaktion auf eine Reihe von Sensoren, mit denen jeder *Tisch* verbunden ist, eine bestimmte Bandbreite von Verhaltensweisen.

Tisch III (Funktionale Beschreibung)

Auf *Tisch III* befinden sich zwei Bildprojektionssysteme: ein spezieller Diaprojektor und ein Videoprojektor. Zu jedem System gehört eine Gruppe von Bildern, die ein bestimmtes Projektionsenvironment beanspruchen: die Standbilder kommen mit einer Anordnung von sechs Bildschirmen aus, während die bewegten Bilder ein weiteres Projektionsenvironment beanspruchen. Beide Projektionsarten arbeiten mit sechsfacher Strahlenaufsplitterung, so daß die Bilder über sechs Achsen kubischer Vektoren auf die sechs Bildschirme verteilt werden. Die Projektion der Standbilder wird durch kleinere

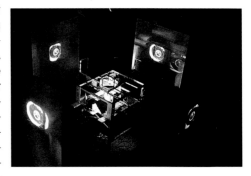

Rahmen unterbrochen, durch die sie ihr eigenes Environment innerhalb des allgemeinen Raumes definiert. Wenn diese kleinen Rahmen verschwinden, ist der Projektionspfad frei für die Sequenz der bewegten Bilder. Dieses erweiterte Projektionsenvironment ist durch die Anordnung von sechs Bildwänden definiert, vier stehen am

Boden, eine hängt von der Decke herab. Das Material der Bildwände ist so beschaffen, daß die Bilder von beiden Seiten sichtbar sind, d.h. daß der Betrachtungsmodus der Installation nach außen hin erweitert wird. So wird die Installation zu einem Objekt aus einer Vielzahl untereinander zusammenhängender Bilder. Dazu kommen weitere funktionale Klang- und Interaktivitätselemente, durch die die Präsenz der Zuschauer und ein gewisses Maß an Partizipation ermöglicht werden.

Auszüge aus einem Gespräch zwischen Woody Vasulka und David Dunn

WV: Ich habe in dieses Stück sehr viele militärische Gegenstände eingebaut. Wenn du das Schild auf diesem Tisch liest, da steht: „Case and Rack Assembly Bomb Navigational Control". Es ist verrückt, daß ich diese Dinge bei mir zu Hause habe, deshalb habe ich das hier in mein Zielsystem integriert, weil es diese Funktion in Wirklichkeit ja auch hatte. Es sollte Bomben steuern, also verwende ich es zur Steuerung meiner Bildkorridore, die im Grunde ja auch Bahnen unsichtbarer Geschoße sind.
DD: Es besteht also ein offener Zusammenhang zwischen der Idee der *Brotherhood* und der Kriegsmaschinerie.
WV: Ich habe nichts dagegen, darüber zu sprechen, denn ich habe mich, während ich intellektuell immer dagegen war, mit diesen Kriegsgeräten umgeben und habe sie in gewisser Weise angenommen. Der RPT-Roboterkopf in *The Theater of Hybrid Automata* ist aus einem Navigationsgerät gemacht, das für den Strategic Air Command Bomber steuerte. Als ich es nach Europa mitbrachte und einem Kollegen in Brünn zeigte, sah er es sich an und sagte: „Jetzt weiß ich, was du machst, ich war nämlich beim ägyptischen Militär Berater für Raketensteuerungssyteme." Er hat nicht nur die *Brotherhood* anerkannt, sondern ist sogar ein „Bruder" der *Brotherhood* geworden.
DD: Für dich wird das also inhaltlich klar. Du hast jahrelang Reste aus Los Alamos als Arbeitsmaterial in deinem Studio eingesetzt.
WV: Jetzt ist es so nackt wie sein Inhalt.
DD: Es ist sicher deutlich, insofern als diese Materialreste der Schutt dieser ganzen Kriegskultur sind. Viele Künstler haben ja die Schrottplätze in Los Alamos durchforstet, um mit dem Material metaphorisch eine Kritik an der Verbindung von Wissenschaft und Militärkultur zu formulieren. Du nimmst ganz spezielle weggeworfene Materialien, modelst sie aber nicht zu einem künstlerischen Objekt um, das dann etwas aussagt, sondern belebst die strukturellen Intentionen dieser Gegenstände als Art reine Artikulation ihrer generativen Ideologie.
WV: Das Ziel bleibt genau dasselbe, nämlich den Geist ihrer Schöpfer zu artikulieren; die männliche Idee der destruktiven Macht der Maschine. Dieses Ding hier, eine rudimentäre Bombenaufhängevorrichtung, ist noch immer von diesem Geist beseelt. Als ich es zum erstenmal

Excerpt from a discussion between Woody Vasulka and David Dunn

WV: I've incorporated vast amounts of military equipment into this piece. If you read the label on this table it's called: "Case and Rack Assembly Bomb Navigational Control." It's crazy that these things come to my house so I took this and incorporated it into my targeting system because this is what it really is. It was designed to navigate bombs so I'm using it to navigate my pictorial corridors which are basically trajectories of invisible projectiles.
DD: So that's an overt connection to this idea of *Brotherhood* and the machinery of war.
WV: I don't hesitate to speak about it because while I have always been intellectually opposed to it, in fact I've surrounded myself with these war machines and have adopted them. In fact the RPT robotic head in *The Theater of Hybrid Automata* is made from a celestial navigation unit that navigated the bombers for the Strategic Air Command. When I brought it to Europe and showed it to one of my colleagues in Brno, he looked at it and said: "now I know what you are doing because I was an adviser to the Egyptian military about missile navigation systems." He not only recognized the *Brotherhood* but became a "brother" of the *Brotherhood*.
DD: So, in your mind, this is becoming explicit as content. For years you have been working with surplus from Los Alamos but it was media related as appropriated materials for your studio.
WV: Now its become very naked as the content itself.
DD: It's certainly upfront in terms of this surplus material being the detritus of that culture of war. Artists here have been raiding the Los Alamos scrap yards in order to make these metaphoric expressions as a kind of critique of the nexus of science and military cultures. But what you are doing is taking very specific cast-off materials and, rather than refashioning them into a sculptural expression, resuscitating the structural intentions of these devices as a kind of pure articulation of their generative ideology.
WV: It has exactly the same purpose, to amplify the mind of its creator: the male idea of the machine's destructive power. This thing, a vestigial bombing rack, carries the inspiration with it. When I saw it for the first time, I knew exactly that this was a piece of that soul. I didn't even know what it was until I read it later but I understood it intuitively. When I opened the box, there

was a table with four legs and these racks which I later read were part of these bombing computers. I envisioned these guys sitting in the jungle, just before they went to Cambodia, programming these computers. They were probably dressed in fatigues, drinking beer, punching the code into computers mounted on these racks. So I'm trying to replicate exactly the spirit contained within this piece of metal. It is probably subconscious but very authentic: these were the machines for automatic bombing so that no one had to have the consciousness or responsibility of inflicting death. These codes are hidden to the general art strategies unless one descends to this level of intimacy where you recognize by strange instinct the role of these objects. I think it transfers subconsciously to the mind of the observer. It is this third level of involvement that really interests me rather than the obvious one.

sah, wußte ich genau, daß das ein Stück dieser Seele war. Ich wußte nicht einmal, was es war, bis ich es später las, aber ich verstand es sofort intuitiv. Als ich die Kiste aufmachte, war ein Tisch drin, mit vier Beinen und diesen Vorrichtungen, die, wie ich später las, zu diesen Bombencomputern gehörten. Ich stellte mir vor, wie die Burschen im Dschungel saßen, kurz bevor sie nach Kambodscha gingen, und die Computer programmierten. Sie hatten vermutlich eine Uniform an, tranken Bier, hackten den Code in die Computer, die auf diesen Gestellen montiert waren. Ich versuche also, genau nachzuvollziehen, welcher Geist in diesem Stück Metall steckt. Das ist wahrscheinlich ein unbewußter, aber gerade deshalb ein sehr authentischer Prozeß: das waren automatische Bomben, so daß niemand sich mit dem Bewußtsein oder der Verantwortung auseinandersetzen mußte, daß er tötete. Diese Codes bleiben in der allgemeinen Kunststrategie im Verborgenen, wenn man sich nicht auf diese Stufe der Intimität hinabbegibt, wo man durch einen seltsamen Instinkt die Rolle dieser Objekte erkennt. Ich glaube, das überträgt sich unbewußt auch auf den Betrachter. Es ist diese dritte Ebene des Beteiligtseins, die mich wirklich interessiert, nicht die, die offensichtlich ist.

NETZHAUT
network skin

Ein Fassadenkonzept für das Ars Electronica Center Linz
a facade concept for the ars electronica center, linz

CHRISTIAN MÖLLER / JOACHIM SAUTER

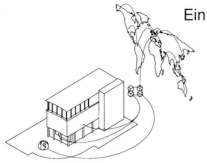

Konzepterläuterungen und Beschreibung der Komponenten:

1.0 Netzwerke, Programme und Daten:
1.1 Globales Netz

Alle an weltweite Computernetze angeschlossenen Benutzer werden über verschiedene Internet-Kanäle, News, E-mail, WWW ... aufgefordert, 3D-Objektdaten (Programm 1) oder 2D-Bilddaten (Programm 2) über Netz an das AEC zu schicken.

1.2 Internes AEC Rechnernetz:

Programm 1: Das AEC Rechnernetzwerk empfängt und sammelt diese Objektdaten. Es plaziert sie automatisch entsprechend ihrer lokalen Herkunft auf einer virtuellen Weltkugel.
Diese künstliche Welt ist duch das Empfangen immer neuer Daten einem dauernden Wandel unterworfen. Bei alten Daten wird, falls die Darstellungskapazität nicht mehr ausreicht, die Geometrie entsprechend ihrem Alter vereinfacht – neues wächst somit auf dem vergänglichen alten.
Prgramm 2: Hierbei werden nicht 3D sondern 2D Daten (statische und bewegte Bilder) als virtuelle Repräsentanten der Absender auf der virtuellen Weltkugel angeordnet. Entsprechend ihrer Herkunft werden sie auf senkrecht zur Weltkugel stehenden Polygonen gemapt. Diese wenden sich beim Überfliegen oder Durchschreiten dem Betrachter zu. („Billboards"). In der virtuellen Welt erhalten diese Objekte ein Verhalten. (Sie verändern sich, verei-

Explanation of the concept and description of the components:

1.0 Networks, programs and data:
1.1. Global network

All users connected to worldwide computer networks will be requested via Internet channels, News, e-mail, WWW ... to send 3D objects data (Program 1) or 2D graphics data (Program 2) to the AEC.

1.2. Internal AEC computer network

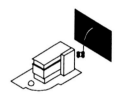

Program 1: The AEC computer network receives and collects these data. It automatically places them on a virtual globe according to their origin. This artificial world is subject to constant change due to its reception of new data. In the case of old data, the geometry is simplified according to the age in the event that the display capacity is no longer sufficient – new data therefore grows on top of the ephemeral old data.
Program 2: In this program, 2D data (static and movable images) rather than 3D data are arranged on the virtual globe as virtual representatives of the sender. According to their origin, they will be mapped on polygons which stand perpendicular to the globe. These polygons turn towards the viewer when he or she makes a walkthrough ("billboards"). These objects are assigned a specific behavior in this virtual world. (They change, unite, etc. according to predefined rules.) The objects are in contact with their senders, whom they represent virtually: When the data is first placed on the globe and at every change, the sender will receive a generated picture of his or her repre-

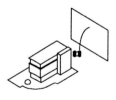

sentative and its environment via the network.

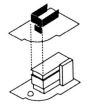

2.0 Displays

A double-sided video wall is located on the ground floor. The network projected onto it can be viewed from the inside and the outside at the same time.

The surface of the building's continuous two-story facade section ("cummerbund") consists of large-sized, etched glass plates. During the day, the "network skin" facade appears to be a closed balustrade element of opaque white glass. In the evening and at night, the opaque glass sheathing functions as the network world's rear projection surface for the outside world.

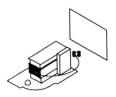

2.1 Southern facade

On the second and third floors of the building's gabled end, there are breaks in the continuous, solid outer wall. Images are projected through large window openings onto the opaque outer sheathing.

In this way, the slight curve in the southern facade incorporates the curve of the projected globe into its form. With regard to the content, the virtual architecture therefore enters into a dialog with the real architecture through this projection.

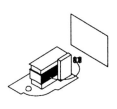

2.2 Eastern and western facades

The metaphorical transfer of the network idea conceived on the two large square surfaces of the eastern and western facades is the dominating formal element of the facade design. From the exterior, the material and construction of the longitudinal facade sections are identical to those of the southern facade elements. In contrast to the former, however, an image is not projected onto the facade, but along the surface of the glass.

Laser beams are projected parallel to the outer

nigen sich etc. nach vordefinierten Gesetzmäsigkeiten). Die Objekte stehen mit ihren Absendern, die sie virtuell repräsentieren, in Verbindung: Bei der ersten Plazierung sowie bei jeder Veränderung erhält der Absender ein gerendertes Bild seiner Repräsentanz und deren Umgebung über das Netz zugesandt.

2.0 Displays

Im Erdgeschoß befindet sich eine doppelseitige, datenfähige Videowand. Die darauf übertragene Netzwelt ist gleichzeitig von innen und außen zu betrachten.

Der umlaufende, zweigeschossige Fassadenteil („Bauchbinde") des Gebäudes besteht in seiner Oberfläche aus großformatigen, geätzten Glastafeln. Bei Tag erscheint die Netzhautfassade als ein geschlossenes Brüstungselement aus weiß-opakem Glas. In den Abend- und Nachtstunden dient die opake Glasverkleidung als Rückprojektionsfläche zur Übertragung der Netzwelt nach außen.

2.1 Südfassade

Im 1. und 2. OG der Giebelseite des Gebäudes ist die umlaufende, massiv ausgeführte Außenwand unterbrochen. Durch großflächige Fensteröffnungen wird von innen auf die opake Außenverkleidung projiziert.

Formal nimmt so die Krümmung der projizierten Weltkugel die leichte Krümmung der Südfassade auf. Inhaltlich tritt durch diese Projektion die virtuelle Architektur in den Dialog mit der realen Architektur.

2.2. Ost- und Westfassade

Die auf den beiden großformatigen Rechtecksflächen der Ost- und Westfassade konzipierte metaphorische Übertragung der Netzidee ist das formal dominante Element der Fassadengestaltung. Von außen entsprechen die längsgerichteten Fassadenteile in Material und Konstruktion genau der Südfassade. Im Unterschied zur Südfassade findet hier die Bildprojektion nicht auf, sondern entlang der Glasebene statt.

Über bewegliche, von den Betrachtern

skin via movable tilted mirrors controlled interactively by the viewers, and these beams trace through the opaque glass of the balustrade element in the form of a whitish green pattern of lines. A strip along the entire edge of the balustrade element which consists of a surface mirror directed towards the interior reflects the beams striking it until they are lost in infinity due to the gradual decrease in their luminous intensity. The lower the angle of projection (which can changed interactively), the denser the lined image on the facade, which reminds one of a wire frame drawing. And so, an interactive, quickly changeable facade texture of varying line density is created.

3.0 Interface:
3.1 Earthtracker:

The Earthtracker is the interface through which the passers-by and visitors can interact directly with the facade. With the aid of a globe which is set into the floor in the same way as a trackball, the projected virtual globe can be moved, thereby allowing the visitors to navigate through it. The real globe is at the same time connected in direct proportion to the virtual one. In addition to the possibilities and direction of movement, which the visitor can control by rotating the globe, the height and direction of the line of sight and can be changed by means of a bow which resembles a latitudinal line (by pulling it up or pushing it down and turning it).

In addition to the network world projected onto the southern side, the Earthtracker also influences the angle of the system of mirrors on the eastern and western facades. This means that all facade sections are controlled in harmony.

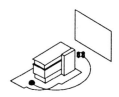

3.2 Network Traffic Visualizer

If the Earthtracker is motionless for an extended period of time, the laser/mirror system will be controlled according to the load or fluctuations of the AEC network traffic.

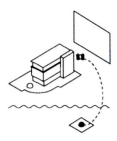

3.3 Interfaces from other locations

At night, the laser network visualization will be visible not only in the immediate surroundings, but also at greater distances from locations with a clear view of the AEC. Other Earthtrackers will be placed at these locations to teleinteract with the AEC's network skin.

3.3 Interfaces von anderen Orten:

Die Lasernetzvisualisierung ist nachts nicht nur in der direkten Umgebung, sondern auch von entfernten Orten mit freiem Blick auf das AEC zu sehen. An diesen Stellen stehen weitere Earthtracker, um mit der Netzhaut des AECs zu teleinteragieren.

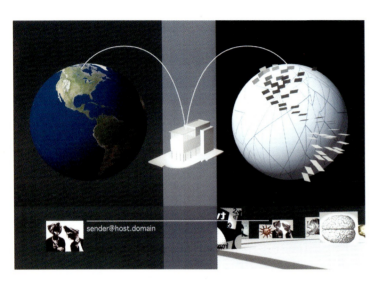

CYBER CITY FLIGHT

Ein virtueller Spaziergang durch Berlin
a virtual stroll through berlin

ART + COM

Das Zentrum Berlins mit seinen geplanten Neubauten für die Hauptstadt- und Regierungsfunktion wird von ART + COM in einem interaktiven Szenario präsentiert. In der Konfiguration „Cyber City Berlin" ist es mit schnellen Grafikworkstations möglich, den aktuellen Planungsstand der Regierungs- und Parlamentsbauten in einem digitalen Stadtmodell zu simulieren. In dieser virtuellen Realität kann der Betrachter – im Gegensatz zu üblichen Architekturanimationen – jeden gewünschten Standpunkt einnehmen und sich nach seinen Vorstellungen durch die Stadt bewegen.

Darüber hinaus hat er die Möglichkeit, Entwürfe gegeneinander auszutauschen und in Gebäude hineinzugehen; das Reichstagsgebäude wurde nach den Umbauplänen von Sir Norman Foster in den Computer eingegeben und kann interaktiv begangen werden. Ermöglicht wird dies durch eine Level-of-Detail-Technik, die bei Annäherung an ein Gebäude immer mehr Details sichtbar werden läßt.

Der Cyber City Flight schafft eine sinnliche Erlebniswelt für Stadträumliche Zusammenhänge und Veränderungen.

The center of Berlin and the new buildings planned for its function as capital and seat of the government will be presented by ART + COM in an interactive scenario. In the configuration "Cyber City Berlin", fast graphic workstations will make it possible to simulate the current planning status of the government administration and parliamentary buildings in a digital model. In this virtual reality, the viewer – in contrast to conventional architectural animations – can assume any position and move through the city as desired.

In addition, exchanging designs and entering the buildings will also be possible; the Reichstag (parliament) building was entered into the computer according to the reconstruction plans of Sir Norman Foster, and the user can enter it interactively. All this was made possible by a level-of-detail technique, according to which more and more details become visible as one approaches a building.

The Cyber City Flight creates a sensory world for experiencing correlations and changes with regard to urban space.

Das digitale Stadtmodell der Berliner Innenstadt

ZEITPLAN
schedule

Monday, June 21

6:00 – 7:00	OPENING OF THE EXHIBITION "THE ARS ELECTRONICA", O.Ö. Landesmuseum
9:00 – 12:00	ELECTRONIC PARTY, Hauptplatz

Tuesday, June 21

10:00 – 1:00	"INTELLIGENT PRODUCTS" SYMPOSIUM Brucknerhaus, Stiftersaal
2:30 – 6:00	"LIFE IN THE NETWORK" SYMPOSIUM Brucknerhaus, Stiftersaal
6:00 – 7:00	OPENING OF THE "INTELLIGENT AMBIENCES" EXHIBITION Design Center
9:00 – 10:00	JARON LANIER, "CYBER INSTRUMENTS" Brucknerhaus, Brucknersaal

Wednesday, June 22

10:00 – 1:00	"ARCHITECTURE AND ELECTRONIC MEDIA" SYMPOSIUM Brucknerhaus, Stiftersaal
2:30 – 6:00	"ARCHITECTURE AND ELECTRONIC MEDIA" SYMPOSIUM Brucknerhaus, Stiftersaal
7:30	ELECTRONIC MUSIC LUDGER BRÜMMER, GÜNTHER RABL
9:00	SOLDIER STRING QUARTET & ELLIOTT SHARP Brucknerhaus, Stiftersaal
9:00 – 10:00	PRIX ARS ELECTRONICA GALA ORF Landesstudio OÖ
10:00	APRÈS ARS – ERLEBNISHAUS K4 Pig roast

Thursday, June 23

10:00 – 6:00	PRIX ARS ELECTRONICA ARTISTS' FORUM, ORF Landesstudio OÖ
8:00 – 9:00	ELECTRO CLIPS C. MÖLLER/ST. GALLOWAY Brucknerhaus, Brucknersaal

Montag, 20. Juni

18.00 – 19.00	ERÖFFNUNG DER AUSSTELLUNG „DIE ARS ELECTRONICA", O.Ö. Landesmuseum
21.00 – 24.00	ELEKTRONISCHES FEST, Hauptplatz

Dienstag, 21. Juni

10.00 – 13.00	SYMPOSIUM „INTELLIGENTE PRODUKTE" Brucknerhaus, Stiftersaal
14.30-18.00	SYMPOSIUM „LEBEN IM NETZ" Brucknerhaus, Stiftersaal
18.00 – 19.00	ERÖFFNUNG DER AUSSTELLUNG „INTELLIGENTE AMBIENTE" Design Center
21.00 – 22.00	JARON LANIER „CYBER INSTRUMENTS" Brucknerhaus, Brucknersaal

Mittwoch, 22. Juni

10.00 – 13.00	SYMPOSIUM „ARCHITEKTUR UND ELEKTRONISCHE MEDIEN" Brucknerhaus, Stiftersaal
14.30 – 18.00	SYMPOSIUM „ARCHITEKTUR UND ELEKTRONISCHE MEDIEN" Brucknerhaus, Stiftersaal
19.30	ELEKTRONISCHE MUSIK LUDGER BRÜMMER, GÜNTHER RABL
21.00	SOLDIER STRING QUARTET & ELLIOTT SHARP Brucknerhaus, Stiftersaal
21.00 – 22.00	PRIX ARS ELECTRONICA GALA ORF Landesstudio OÖ
22.00	APRÈS ARS – ERLEBNISHAUS K4 Spanferkelei

Donnerstag, 23. Juni

10.00 – 18.00	PRIX ARS ELECTRONICA KÜNSTLERFORUM ORF Landesstudio OÖ
20.00 – 21.00	ELECTRO CLIPS CHRISTIAN MÖLLER/STEPHEN GALLOWAY Brucknerhaus, Brucknersaal

| 21.00 – 22.00 | SEVEN GATES, MARK TRAYLE
Brucknerhaus, Brucknersaal |
| 22.00 | APRÈS ARS – ERLEBNISHAUS K4
Konzert für Bomben & Granaten |

Freitag, 24. Juni

| 10.00 – 18.00 | PRIX ARS ELECTRONICA KÜNSTLERFORUM
ORF Landesstudio OÖ |
| 14.30 – 17.00 | ABSCHLUßPRÄSENTATION „COMPUTER UND SPIELE"
Brucknerhaus, Stiftersaal |
| 20.00 – 24.00 | AMBIENCE PERFORMANCE
Brucknerhaus, Brucknersaal |
| 24.00 | APRÈS ARS – ERLEBNISHAUS K4
Techno für Fortgeschrittene |

Samstag, 25. Juni

| 21.30 | FIESTA ELECTRA, FRIEDRICH GULDA AND HIS PARADISE BAND
Sporthalle (open end) |

Laufende Veranstaltungen:

INTELLIGENTE AMBIENTE, Ausstellung, Design Center, 21.- 26.Juni, täglich 10.00 – 19.00 Uhr

DIE ARS ELECTRONICA, Ausstellung, OÖ. Landesmuseum Francisco Carolinum, 21.Juni – 10.Juli, Montag – Samstag, 10.00 – 18.00 Uhr

INTELLIGENTE ENVIRONMENTS, Installationen, Brucknerhaus, Foyer, 21. – 24. Juni, 9.00 – 21.00 Uhr, 25. Juni, 9.00 – 18.00 Uhr

INTELLIGENT AMBIENCE, Videoprogramm, Brucknerhaus, Keplersaal, 21.- 24.Juni, 10.00 – 21.00 Uhr, 25. Juni, 10.00 – 18.00 Uhr

PIAZZA VIRTUALE: SERVICE AREA A.I., auf Internet: 24 Stunden; Brucknerhaus: 9.00 – 21.00 Uhr, Samstag, 9.00 – 21.00 Uhr, 21.- 25.Juni

COMPUTER UND SPIELE, Interaktive Ausstellung von Schülerarbeiten, Brucknerhaus, Foyer, 21. Juni – 25. Juni, 10.00 – 20.00 Uhr

PRIX ARS ELECTRONICA
Videothek – Audiothek – Ausstellung, ORF Landesstudio OÖ, 23.-25.Juni, 10.00 – 18.00 Uhr

APRÈS ARS – ERLEBNISHAUS K4, Stadtwerkstatt, Kirchengasse 4, 20.Juni – 25.Juni. Gastgarten & Cafe Bar STROM ab 18 Uhr, Cocktail Lounge ab 21 Uhr

MIT DEN AUGEN DER ARCHITEKTUR, Ausstellung, Offenes Kulturhaus, Dametzstr. 30, 21.Juni – 29.Juli, täglich 14.00 – 20.00 Uhr

PUBLIC INTERVENTION, Galerie MAERZ, Landstraße 7 8.-29. Juni, Mo-Fr. 15,00 – 18.00 Uhr, Sa 10.00 – 13.00 Uhr

| 9:00 – 10:00 | SEVEN GATES, MARK TRAYLE
Brucknerhaus, Brucknersaal |
| 10:00 | APR'ES ARS – ERLEBNISHAUS,
Concert for Bombs and Grenades |

Friday, June 24

| 10:00 – 6:00 | PRIX ARS ELECTRONICA ARTISTS' FORUM, ORF Landesstudio OÖ |
| 2:30 – 5:00 | CLOSING PRESENTATION, "COMPUTER AND GAMES"
Brucknerhaus, Stiftersaal |
| 10:00 – 12:00 | AMBIENCE PERFORMANCE
Brucknerhaus, Brucknersaal |
| 12:00 | APRÈS ARS – ERLEBNISHAUS K4
Techno for the advanced |

Saturday, June 25

| 9:30 | FIESTA ELECTRA, FRIEDRICH GULDA AND HIS PARADISE BAND
Sporthalle (open end) |

Continous events:

INTELLIGENT ENVIRONMENT, exhibition, Design Center, June 21 – 26, daily, 10:00 AM – 7:00 PM

THE ARS ELECTRONICA, exhibition, OÖ Landesmuseum Francisco Carolinum, June 21 – July 10, Mon. – Sat., 10:00 AM – 6:00 PM

INTELLIGENT ENVIROMENTS, installations, Brucknerhaus, Foyer, June 21 – 24, 9:00 AM – 9:00 PM, June 25, 9:00 AM – 6:00 PM

INTELLIGENT AMBIENCE, video program, Brucknerhaus, Keplersaal, June 21 – 24, 10:00 AM – 9:00 PM, June 25, 10:00 AM – 6:00 PM

PIAZZA VIRTUALE: SERVICE AREA A.I., on Internet: 24 hours; Brucknerhaus: 9:00 AM – 9:00 PM, Saturday, 9:00 – 9:00 PM June 21 – 25

COMPUTER AND GAMES, Interactive exhibition of students' works. Brucknerhaus, Foyer, June 21 – 25, 10:00 AM – 8:00 PM

PRIX ARS ELECTRONICA
Videothek – Audiothek – Exhibition, ORF Landesstudio OÖ, June 23 – 25, 10:00 AM – 6:00 PM

APRÈS ARS – ERLEBNISHAUS K4, Stadtwerkstatt, Kirchengasse 4, June 21 – 25. STROM outdoor café and bar, opens at 6:00 PM, Cocktail lounge after 9:00 PM

WITH THE EYES OF ARCHITECTURE, exhibition, Offenes Kulturhaus, Dametzstraße 30, June 21 – July 29. Daily, 2:00 PM – 8:00 PM

PUBLIC INTERVENTION, Galerie MAERZ, Landstraße 7, June 8 – 29, Mo – FR, 3:00 PM – 6:00 PM, Sa 8:00 AM – 1:00 PM

INTELLIGENTE PRODUKTE

Technologie für Menschen mit Behinderungen

„Der Behinderte, der seit langem mit Hilfe von technischen Prothesen lebt, die seine fehlenden Funktionen ersetzen, wird zu einer Modellfigur, die neues Licht auf das Ziel der technischen Evolution wirft."

PETER WEIBEL

1.
FACHWERK Arbeitsplätze für Menschen mit Behinderungen stellt bei Ars Electronica 94 das „Telephone Subset For Disabled People (TSFDP)" vor. Das TSFDP kann die Kommunikation über Telefon solchen Menschen erleichtern, die an der Parkinsonschen Krankheit leiden, Menschen, die an Muskellähmung erkrankt sind, alten Menschen, bewegungsbeeinträchtigten und bewegungsunfähigen Menschen und allen Personen, die kleine Tasten oder Schalter nicht betätigen können. Mit einem kurzen Pfiff des Benutzers wird ein „Zeitfenster" geöffnet, während dessen Dauer es möglich ist, durch weitere kurze Pfiffe eine von drei vorher abgespeicherten Nummern zu selektieren. Wenn sich der Gesprächspartner meldet, kann sofort über eine Freisprecheinrichtung gesprochen werden. Auch beim Läuten des Telefons erfolgt das „Abheben" des Hörers mit einem kurzen Pfiff. Das TSFDP wird derzeit in einer verbesserten Form weiterentwickelt, um z.B. mit einem Computer ferngesteuert werden zu können. Informationen: FACHWERK Arbeitsplätze für Menschen mit Behinderungen, Braunhubergasse 4A, A-1110 Wien.

2.
CARETEC Technische Hilfsmittel für Behinderte GesellschaftmbH stellt die folgenden Produkte aus:
– COLORTEST Farberkennungsgerät für Blinde und Farbenblinde: Dieses Gerät vermittelt durch Spekralanalyse folgende Werte von Objekten: Farbton, Sättigungsgrad, Helligkeit und prozentuelle Zusammensetzung einer Farbe. Die Ergebnisse der Analyse werden in gut verständlicher deutscher Sprache wiedergegeben. Das Gerät selbst ist leicht, klein und handlich und kann universell eingesetzt werden.
– LUMITEST zur Lichterkennung, Helligkeits- und Kontrastmessung und zum Aufspüren von Lichtquellen.
– NOTAPHON Notizgerät mit Blindentatstatur und Sprachausgabe. Das elektronische Notebook kann gespeicherte Texte auf einer angeschlossenen Braillezeile lesegerecht darstellen und mit einem angeschlossenen Braille- oder Schwarzschriftdrucker ausdrucken. Es ist für blinde, sehbehinderte und sehende Menschen gleichermaßen verwendbar.
– Das sprechende Fieberthermometer CIBERVEU mißt die Körpertemperatur elektronisch und gibt sie in deutlicher synthetischer Sprache aus.
Informationen: CARETEC Österreich,
Wollzeile 9, A-1010 Wien

3.
IBM Österreich präsentiert in Zusammenarbeit mit der APA (Austria Presse Agentur) eine „elektronische" Zeitung. Die APA bietet seit einiger Zeit die Möglichkeit, über Telefonleitung und Modem die täglichen Ausgaben einiger österreichischer und ausländischer Zeitungen direkt und aktuell auf dem Bildschirm des Heim- oder Bürocomputers empfangen zu können. Über das IBM Sprachausgabesystem können auch Blinde und Sehbehinderte die tägliche Zeitung ihrer Wahl „lesen" und darüber hinaus selektiv Artikel und Themenbereiche auswählen.

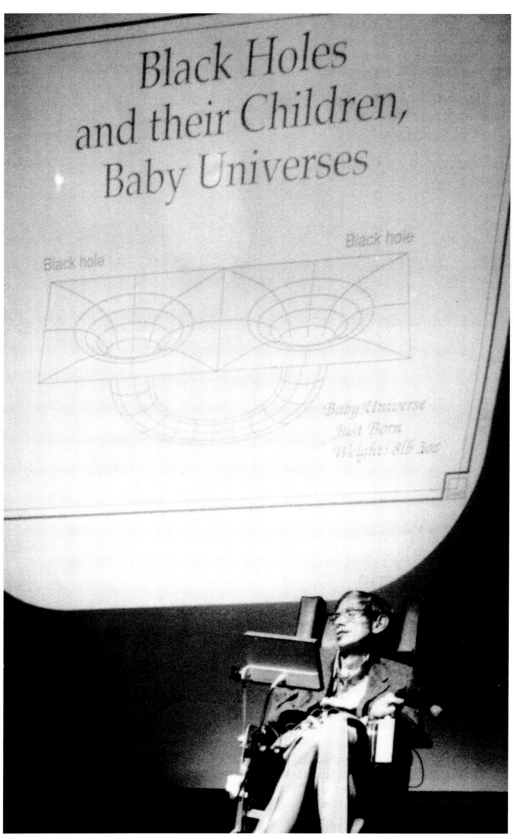

Der Astrophysiker Stephen Hawking

CYBER
MUSIC

ALLES SPIEL

audience participation

LOREN & RACHEL CARPENTER
CINEMATRIX™

Introduction

The patented CINEMATRIX™ Interactive Entertainment Systems gives each person in an audience the real time capability to interact with images on a screen and thus play games together, answer questions, create patterns and make decisions about adventures in a collective virtual world. This invention grew from the innovative world of computer graphics along with advances in computer hardware enabling new forms of interactivity to arise. Up to now, this system has always been done indoors with people in their seats. We propose a unique experience for Linz, to set up the experience outdoors for a standing crowd. We also intend to push the boundaries of what is possible in the experience itself. Necessarily, the games have been fairly simple up until now. We think that it is possible for groups to interact and cooperate to create forms using genetic algorithms. Linz is the perfect place to try this out.

Communication with Humans

Communication is unpredictable at the best of times. The more people you try to communicate with just accentuates the possibilities for misunderstanding and confusion. Having a few calibrating exercises at the beginning helps people to know that the system is working and that they are being represented. Before giving explanations, it is important for the audience to experiment with the wands, this can produce some fun surprises. Then the group can be asked to show one color and they can see it up on the screen. When the group changes the color they are showing, it changes correspondingly on the screen. The participants can then be divided into groups and reproduce the exercise to get an even better idea of where their position is on the screen and to get a feeling for the system working.

Einführung

CINEMATRIX™ Interactive Entertainment Systems, eine inzwischen patentierte Technologie, schafft vollkommen neuartige Interaktivitätsformen. So können Zuschauer individuell und in Echtzeit mit Bildern auf einem Bildschirm interagieren und miteinander Spiele spielen, Fragen beantworten, Muster erzeugen und Entscheidungen über Abenteuer in einer kollektiven virtuellen Welt treffen. Möglich wurde diese Erfindung durch Innovationen in der Computergrafik und auf dem Hardware-Sektor. Bisher haben wir das System immer in einem geschlossenen Raum eingesetzt, in dem die Zuschauer saßen. In Linz wollen wir ein neuartiges Erlebnis schaffen, und zwar im Freien mit einer stehenden Zuschauermenge. Wir wollen die Grenzen dessen, was mit dem System möglich ist, noch erweitern. Bisher waren die Spiele notgedrungen relativ einfach. Wir sind aber davon überzeugt, daß mit Hilfe genetischer Algorithmen Gruppen interaktiv und kooperierend Formen kreieren können. Ars Electronica ist für ihre Innovationsfreudigkeit bekannt, und wir glauben deshalb, daß Linz der ideale Ort für die Premiere dieses kinoetischen Evolutionsschrittes ist.

Menschliche Kommunikation

Kommunikation ist bestenfalls unvorhersagbar. Mit je mehr Menschen man zu kommunizieren versucht, desto zahlreicher sind die Möglichkeiten für Mißverständnisse und Verwirrung. Ein paar Einführungsübungen am Anfang haben den Zweck, den Zuschauern zu zeigen, daß das System funktioniert und sie darin ihren Platz haben. Wichtig ist auch, daß das Publikum, bevor man mit Erklärungen beginnt, erst einmal mit den Zauberstäben experimentiert; das kann zu erheiternden Überraschungen führen. Dann kann man das Publikum z.B. auffordern, eine bestimmte Farbe anzuzeigen, die es dann auf der Bildwand sieht. Wenn eine andere Farbe angezeigt wird, ändert sich auch die Farbe auf der Bildwand entsprechend. Wenn man dann das Publikum in mehrere Gruppen teilt und die Übung wiederholen läßt, bekommen die einzelnen Teilnehmer eine bessere Vorstellung davon, wo ihre jeweilige Position auf der Bildwand ist, und können die Funktionsweise des Systems besser verstehen.

Die Verwendung passiver Reflektoren als Interface macht eine Verkabelung überflüssig und ermöglicht ein unmittelbares Feedback. Die Teilnehmer können agieren, ohne kompliziert Tasten drücken oder Hebel betätigen zu müssen. Das Konzept des Systems ist auf diese Art nicht nur insgesamt einfacher, sondern auch leichter nachvollziehbar. Kein Publikum hört sich gerne komplizierte Regeln an. Die Spiele müssen visuell verständlich sein und die richtige Mischung aus aktiver Teilnahme und verlockenden Aufgabenstellungen anbieten.

Hardware- & Software-Anforderungen

Das ursprüngliche Event mit Publikumspartizipation fand nach einer Entwicklungs- und Vorbereitungszeit von nur fünf Monaten im Rahmen der SIGGRAPH '91 statt. Viele Fragen mußten beantwortet und viele Probleme gelöst werden. Das erste große Problem war das Interface. Wie konnte das Publikum auf billige, sichere Art und Weise Signale an den Computer schicken? Taschenlampen wären zu teuer gewesen, und jeden Sitz zu verkabeln, kam nicht in Frage. Wie stand es aber mit dem Reflektieren von Licht? So kamen wir auf den sogenannten „Zauberstab", einen einfachen Stab mit „Reflexite".
Die Reflektoren sind die erste Beschränkung beim Entwurf eines solchen Spiels. Die Positionierung der Zuschauer ist äußerst wichtig. Das reflektierte rote oder

Using passive reflectors as an interface allows the freedom of no wires and immediate feedback. People can respond without complicated button-pushing or the moving of levers. It helps for people to have a familiarity with the concept of the game and for the goal to be simple. Large crowds do not want to listen to a lot of rules. The games need to be visually understandable with the right mix of action and challenge.

Hardware & Software Challenges

The original audience participation event at SIGGRAPH '91 only had five months to invent the system and get it working. Many questions had to get answered and many problems had to be solved. The first big question was interface. How was the audience going to send a signal to the computer in a cost effective, safe way? Flashlights would be too expensive; wiring each seat was out of the question. How about reflecting light? The resulting device of combined "Reflexite" on a paint stick is referred to as a "wand".
The requirements for designing any game with this system begin with the constraints of the

reflective device. Locating the players is a big technical challenge. The red or green reflected light signal needs to be picked up by video cameras, and communicated to the computer, which deciphers the information according to the game or task on the screen. To differentiate each person's signal is a top priority; each person needs to be represented. The camera sees a mass of spots that need to be diferentiated by the computer. Forming a seating chart is one way to partially define where each person is. There are additional considerations of extra movement, reflected light objects in the room. The solution is in software. The past three years of development have made it possible to get away from seating charts when needed. This new method will debut in Linz for ARS Electronica, known for being on the cutting edge of art and technology.

Each venue has its specifications and idiosyncrasies. Basic photographic principles apply. A wide-angle lens on the camera can see more if it has a greater distance. Cameras and lights need to be linked to power with red, green, blue and sync going into the right ports in the computer. It is important that each person be represented.

The Games

It is important to establish that people are having real time input as described in the introduction. Each game requires a small explanation. We try to emphasize games where people work together either in teams or with the entire group. One two-team game has a "paddle" on each side of the screen to hit a conference logo or whatever is appropriate for the event. Green makes the team's paddle move up and red makes it go down. If one side sees that their paddle needs to move up, they start to show the green side of their reflector to the cameras and lights. If everyone on their side showed green, the paddle goes up too far, so people need to moderate the signal by a few people showing red or putting their paddle down. These decisions have to be made quickly and become intuitive producing a state of flow, which ends up being very exciting. New variations of this game are being created continually. The rotating cube is an exercise that requires the entire audience to cooperate. A multicolor-

grüne Lichtsignal muß von Videokameras aufgefangen und an den Computer weitergeleitet werden, der dann die Information je nach Spiel oder gestellter Aufgabe dechiffriert. Dabei ist besonders wichtig, daß das Signal jedes Teilnehmers einzeln erfaßt wird; jeder Teilnehmer muß repräsentiert werden. Die Kamera sieht eine Menge von Punkten, die vom Computer voneinander unterschieden werden müssen. Mit Hilfe eines Sitzplans läßt sich zumindest annähernd die Position jeder Person bestimmen. Dazu kommen dann aber Faktoren wie unvorhersehbare Bewegungen oder reflektierende Gegenstände im Raum. Die Lösung liegt deshalb in der Software. In den letzten drei Jahren gelang es, daß, wenn nötig, auf Sitzpläne verzichtet werden kann. Diese neue Methode wird in Linz im Rahmen von Ars Electronica, die den Grenzbereich von Kunst und Technologie als ihre Herausforderung ansieht, erstmals zum Einsatz kommen.

Jeder Ort hat seine Besonderheiten und Eigentümlichkeiten. Grundsätzlich gelten die Gesetze der Photographie. Mit einem Weitwinkelobjektiv sieht die Kamera mehr; Kameras und Lichter brauchen Strom; Rot, Grün, Blau und Sync müssen in die richtigen Computerports geführt werden. Wichtig ist, daß jeder Teilnehmer repräsentiert wird.

Die Spiele

Wie eingangs erwähnt, erfolgt der Input der Zuschauer in Echtzeit. Jedes Spiel muß kurz erklärt werden, wobei es uns vor allem um Spiele geht, bei denen das Publikum entweder Teams bildet oder als ganzes zusammenarbeitet. Bei einem Spiel für zwei Teams gibt es z.B. auf jeder Seite der Bildwand einen „Schläger", mit dem ein Konferenzlogo oder irgendetwas für die Gelegenheit Passendes bewegt wird. Grün bewegt den Schläger des Teams nach oben, Rot nach unten. Wenn eine Seite sieht, daß ihr Schläger sich nach oben bewegen soll, halten die Teilnehmer die grüne Seite ihres Reflektors in Richtung der Kameras und Lichter. Wenn jeder im Team Grün anzeigt, geht der Schläger zu weit nach oben, und die Teilnehmer müssen das Signal abschwächen, indem ein paar von ihnen Rot anzeigen oder ihren Reflektor nach unten halten. Diese Entscheidungen müssen schnell und intuitiv getroffen werden und führen zu einer fließenden Bewegung, was sehr spannend sein kann. Von diesem Spiel werden ständig neue Varianten entwickelt.

Der rotierende Würfel ist eine Übung, bei der das ganze Publikum zusammenarbeiten muß. Ein bunter Würfel rotiert auf zwei Achsen, wobei jede Achse von einer Hälfte des Publikums gesteuert wird. Rot dreht den Würfel in eine Richtung, Grün in die andere. Das Ziel ist, den

Würfel dann anzuhalten, wenn die blaue Seite sichtbar ist. Das ist keine leichte Aufgabenstellung, aber ein Publikum, das schon aufgewärmt ist, wird keine Schwierigkeiten haben, den Würfel in die richtige Position zu manövrieren.

Ein Flugzeug zu fliegen verlangt von allen Teilnehmer viel Geschick. Eine Gruppe steuert die horizontale, die andere die vertikale Neigung. Dieses Spiel wurde mehrfach modifiziert, um es für die Teilnehmer spannender zu machen.

Das Feedback, das wir auf diese Spiele bekommen haben, hat gezeigt, worauf bei der Entwicklung neuer Spiele besonders zu achten ist. Die Teilnehmer müssen genau wissen, wo sie sind und wie ihr Reflektor mit dem Geschehen zusammenhängt. An der binären Beantwortung von Fragen finden sie nur dann Spaß, wenn sie das Thema wirklich interessiert und es sich um einen Wettbewerb handelt. Spannend wird es für konkurrierende Teams bei solchen Wahr/Falsch-Fragespielen nur, wenn sie das Ergebnis sofort sehen.

Bei Spielen, die von den Teilnehmern nicht Exaktheit, sondern intuitive Entscheidungen verlangen, scheinen Menschen anders miteinander zu kommunizieren und ihr normales, lineares Denkschema abzulegen. Das könnte unter anderem erklären, weshalb so viele Menschen bei diesen Spielen so aufgeregt sind. Das CINEMATRIX™-System gibt ihnen die Gelegenheit, auf eine ganz neue Art miteinander zu arbeiten. Die Technologie ist noch jung und für neue Ideen offen. Von verschiedenen Seiten wurde eingewendet, daß man mit diesem System Menschen, die sich in einem erregten Gemütszustand befinden, steuern könnte. Nachdem es dabei aber vor allem um persönliche Entscheidungen geht, halten wir das für nicht sehr wahrscheinlich, haben das System aber trotzdem in den USA, in der EU, in Japan und in einigen anderen Ländern patentieren lassen. Unsere Technologie soll dazu verwendet werden, Menschen Freude zu bereiten und ihre Kooperationsfähigkeit zu entwickeln.

Solch ein kollaboratives Erlebnis unter Verwendung genetischer Algorithmen wird im Rahmen von Ars Electronica erstmals vorgestellt. Das Publikum wird durch seine Entscheidungen kollektiv einen Organismus erschaffen. Dabei ist das Ergebnis jeder Entscheidung sofort sichtbar. Es können mehrere Segmente gleichzeitig bearbeitet werden, und die Entscheidungen werden sofort umgesetzt. Das ist etwas ganz Neues und wir wissen nicht genau, was passieren wird oder was inzwischen sonst noch erfunden wird. Das macht die ganze Sache so spannend.

ed cube is rotated on two axes, each controlled by one-half of the audience. Red rotates the cube one way and green rotates it the opposite way. The trick is to stop the cube with the blue side showing. It is a difficult problem, but an audience that is warmed up can quickly maneuver the cube into the required position.

Flying an airplane requires very fine skills from the entire group of participants. One side controls pitch, the other yaw. This game has been undergoing rewrites to make it more exciting for groups to play.

The feedback on these games has given us ideas to keep in mind when developing new games. People really need to know where they are and how their reflector relates to the action. Asking questions with binary decisions may produce answers, but that process does not excite the crowd unless they are really interested in the subject matter and it is a competition. Sometimes groups divided up into competing teams can get very excited by true/ false questions, because they can see instant results.

Games that require participants to let go of exactness and make intuitive decisions seem to put people in touch with each other in a different way, to go beyond the usual linear mode of thinking. That could be one of the reasons so many people got excited. The CINEMATRIX system provides the opportunity for people to work together in a way never before possible. It is still a young technology ready for new ideas. Some people have pointed out that there might be a possibility for people to use this in a way to control others when they get in an excited state. Since the entire experience is focused on personal choice, we think this is not likely, but we do own the patent for the U.S, the E.C., Japan and several other countries. It is our intention that the technology be used for people to have fun and develop cooperation in the process.

The genetic algorithm collaborative experience is new for ARS Electronica. The group will get to make choices about the collective creation of an organism. When each choice is made, the crowd will see the results instantly. More than one segment can be worked on at a time and the results will happen quickly. This is new, we don't know exactly what will happen or how many other new things will be invented in the meantime. That's what the cutting edge is all about.

ELEKTRONISCHES FEST

STADTWERKSTATT

On the subject:

Technologically intelligent ambiences are projected as varied functional environments, electronically controlled living spaces which offer a technically perfected environment as a reaction to general needs. An enabled environment which serves the user. Today, a urinal with a light barrier, for example. In the past, pissing without flushing.

A nice, new world; it does what you want it to.

On the surface, one tends to imagine a land of milk and honey; life simply becomes simpler. Annoying manual labor is taken care of by electronics and mechanisms.

The refinement / pulverization of existence

A really intelligent ambience must not let its intelligence lead to the idiocy of its users. An extremely intelligent ambience is not necessarily dependent on technological aids. An electronically controlled ambience must not only lead to the routine facilitation of everyday life; it must also take areas of unexpected stimulus into consideration.

A simpler life in a more complex environment.

The goal is to design the environment's intelligence so that it is conducive to the user's intelligence. However, this also requires that this ambience integrates itself into the big picture into broader contexts. In this sense, everyday processes are put into motion, obstacles are created, and resistances are set, all of which stimulate the user to think and act. A power failure – an example of a learning aid. Friction – a challenge to interact.

Zum Thema:

Technologisch intelligente Ambiente sind projektiert als vielseitige funktionale Umgebungen, elektronisch steuerbare Lebensräume, die, auf allseitige Bedürfnisse reagierend, einen technisch perfektionierten Umraum bieten. Eine befähigte, dem Nutzer dienliche Umgebung. Heute z. B. Pissoir mit Lichtschranke. Einst Pissen ohne Spülen.

Neue, brave Welt, sie tut was man von ihr will.

Vordergründig neigt man dazu, sich ein Schlaraffenland vor Augen zu führen, das Leben wird einfach einfacher. Lästige Handgriffe erledigen sich mittels Elektronik und Mechanik wie von selbst.

Die Verfeinerung/ Pulverisierung des Daseins.

Seiner Intelligenz wegen kann ein wirklich intelligentes Ambiente nicht zur Verblödung der Benutzer führen. Ein außerordentlich intelligentes Ambiente ist nicht unbedingt von technologischen Hilfsmitteln abhängig. Ein elektronisch gesteuertes Ambiente kann nicht ausschließlich zur routinisierten Erleichterung des Alltags führen, sondern muß auch Bereiche der unerwarteten Anregung berücksichtigen.

Einfacher Leben in einem komplexeren Environment

Die Aufgabe ist es, die Intelligenz der Umgebung förderlich für die Intelligenz der Nutzer zu gestalten, vordringlich unter dem Aspekt einer Gesamtschau, einer Gesamtsumme der Wirkungen, z.B. der Vernetzungen etc. In diesem Sinne werden Abläufe des Alltags inszeniert, Hindernisse gebaut und Widerstände vorgegeben, welche die Benutzer zum Denken und Handeln anregen. Stromausfall – ein Beispiel zur Lernhilfe. Reibung – eine Aufforderung zur Interaktion.

Hauptplatz: Beispiel STWST-Eingriffe

Abrieb der Schuhe an der Oberfläche:

Vormals als schönster Platz Europas bezeichnet. 1979 Neugestaltung des Hauptplatzes: Barrieren und Stufen wurden eingebaut, die großzügige Gesamtheit des Platzes wurde durch Gliederungsmaßnahmen zerstört.
Das reibungsreiche, hügelige Kopfsteinpflaster ist reibungsarmer Steinplattenarchitektur und unnützen Stufen gewichen. Durch Reibung an diesem Umbaukonzept des Hauptplatzes ist 1979 die Stadtwerkstatt erstmals öffentlich in Aktion getreten.

Hauptplatz (Main Square): Examples of STWST operations

Scraping off shoes onto the surface:

Previously described as the most beautiful square in Europe. Remodelling of the Hauptplatz in 1979; barriers and stairs are constructed; the square's expanse unity was destroyed by the organizational measures. The bumpy cobblestones, which provide friction, had to make way for frictionless stone slabs and useless steps. Because of the friction caused by this remodelling concept for the Hauptplatz in 1979, the Stadtwerkstatt went into public action for the first time.

Hauptplatz 1920

Abrieb des Komödianten an Öffentlichkeit:

Zum Ausklang des Zeitalters der Performance-Kultur liefert Gerald Wilhelms STWST-Theater einen jetztzeitlichen Beitrag zur Commedia dell'arte, zum öffentlichen Leben am Hauptplatz.

Scraping the comedian off onto the public:

To sound out the age of performance culture, Gerald Wilhelm's STWST-Theater makes a contemporary report on the commedia dell' arte and public life on the Hauptplatz.

Scraping art off onto the construction site of the underground garage:

In 1987, the Hauptplatz was dug up for an underground garage. Only a bunker from the Nazi era stood in the way. STWST staged a two-hour concert at the construction site, "Hochzeit der Bagger" ("Marriage of the Power Shovels"), which integrated the site's machinery and complementary audio installations (Ars. E. 1987).

Abrieb der Kunst an der Großbaustelle der Tiefgarage:

1987 wird der Hauptplatz für den Bau einer Tiefgarage untergraben. Einzig ein Hitlerbunker stellt sich unüberwindbar in den Weg. STWST inszeniert auf der Baustelle unter Einbindung des vorhandenen Baugeräts und mit ergänzenden Klanginstallationen ein zweistündiges Baustellenkonzert (Ars E. 87) „Hochzeit der Bagger".

Hauptplatz 1987

Scraping public's eye on shattering glass:

In 1990, the STWST congratulated the city of Linz on its 500th anniversary: Before the eyes of the public, 4 metric tons of glass were dumped onto the square and shoveled back onto the bed cargo area by city workers. A glistening and clashing symphony of felicitations: "Broken glass brings luck".*

Abrieb des öffentlichen Auges an zersplitterndem Glas:

1990 gratuliert die STWST der Stadt Linz zu ihrem 500jährigen Jubiläum: Vor den Augen der Öffentlichkeit werden vier Tonnen Glas auf den Hauptplatz gekippt und durch öffentliche Körperschaften wieder auf die Ladefläche zurückgeschaufelt. Ein glitzernde und klirrende Glückwunschsinfonie „Scherben bringen Glück".

Hauptplatz 1990

Reibung der Öffentlichkeit an intelligenten Ambientefallen:
Auftakt

„Interaktive Kunst" wird einem breiten Publikum zugänglich und erlebbar gemacht.

1. Öffner

Der Hauptplatz wird in ein quasi-intelligentes Ambiente verwandelt, in dem das Publikum Akteur ist. Jeder Eingang ist so gestaltet, daß der Passant den Eindruck gewinnt, den Platz wie eine Bühne zu betreten. Applaus, Blitzlichtgewitter, auffordernde Zurufe begrüßen den Besucher. Am Platz wird dem Gast automatisch die Möglichkeit eingeräumt, Geschehnisse auszulösen. Fallweise wird die einzelne zur Hauptdarstellerin und im nächsten Moment ist sie Mitglied eines großen Chores.
In diesem Parcours von elektrischen Ereignisauslösern obliegt es dem Engagement des Gastes, ob ihm Gestaltung passiert oder ob er gezielt durch Schaltungen ins Geschehen eingreift und über Schaltungen mit Maschine und Mensch in Interaktion tritt.
Die Dramaturgie des „Öffner" ist so angelegt, daß das „Publikum" zur „Audience Participation" geführt wird.

2. Audience Participation

Ein Laserstrahl kennzeichnet das Spielfeld, in dem das Publikum zur Aktion schreiten wird. Mit zunehmender Dunkelheit wird das Spielfeld immer besser sichtbar. Stadtwerkstatt betreut dieses Projekt organisatorisch, technisch und künstlerisch. Nähere Informationen zum Projekt entnehmen Sie den folgenden Seiten.

Konzeption: THOMAS LEHNER, GEORG RITTER, GOTTHARD WAGNER
Realistion: Dominique Bejvl, Peter Donke, Andi Ehrenberger, Andreas Feichtner, Martina Hufnagl, Andreas Kozmann, Georg Lindorfer, Gitti Vasicek, Marc Vojka, Christine Zigon, in Zusammenarbeit mit dem Ars Team LIVA und dem ORF.

Scraping the public against intelligent ambient traps:
Prelude

"Interactive art" is made accessible to a wide public.

1) Opener

The Hauptplatz will be transformed into a type of intelligent ambience in which the audience are the protagonists. Every entrance is designed in such a way so that the passer-by seems to step onto a stage when entering the square. Applause, a storm of camera flashbulbs, and calls of encouragement greet the visitor. On the square, he or she is automatically given the opportunity to make events happen. The individual could either be the lead actress or, in the next moment, a singer in a large chorus.
In this course composed of electronic triggers of events, the visitor's involvement determines whether things happen to him or her, or whether he or she intentionally intervenes in the events and interacts with man and machine via control switchings.
The scenario of the "Opener" is set up so that the "visitors" are encouraged to engage in audience participation.

2) Audience Participation

A laser beam characterizes the playing field in which the audience will go into action. As night falls, the playing field becomes more and more visible. Stadtwerkstatt is in charge of this project, the organizational, technical and artistic aspects.

Contact: STWST, Kircheng. 4,
A-4020 Linz, Austria
Tel.: +43/732/231209 Fax: +43/732/711846

Conception: THOMAS LEHNER, GEORG RITTER, GOTTHARD WAGNER
Realization: Dominique Bejvl, Peter Donke, Andi Ehrenberger, Andreas Feichtner, Martina Hufnagl, Andreas Kozmann, Georg Lindorfer, Gitti Vasicek, Mark Vojka, Christine Zigon, in cooperation with the Ars Team Liva and the Austrian broadcasting corporation.

* Trans. note: The title of this work is a saying in German which has no correlation in English.

DER SOUND
VON EINER HAND
the sound of one hand

JARON LANIER

I started on the design of the Virtual World for The Sound of One Hand, and on learning how to play it, only a month before the performance, so I had to become completely immersed in the creative process.

I had originally thought the piece would be an elaborate VR 'demo', or explication, with clear visual cues for the music, easy-to-use interfaces, and lots of funny Rube Goldberg tricks. But as I worked on the World, a mood, or an Essence, started to emerge, and it was true to my emotional and spiritual experience at the time. This was unexpected and exciting, even if the content was not cheerful. So I went with a darker and more intuitive process instead of falling in line with the familiar computer culture of clarity and light humor. There have only been rare occasions when I felt I was programming in an intuitive way, and this was one.

Don't expect the instruments to be immediately understandable, or imagine that they are easy to play. They emerged from a creative process I cannot fully explain, and I had to learn to play them. I don't think the two esthetics I'm distinguishing must be mutually exclusive, but the intuitve side of the equation can't reliably be willed into action. A synthesis of clarity and mood will come by grace, when it comes.

The first instrument is called the Rhythm Gimbal. A gimbal is a common mechanical construction; a hierarchy of rotating joints. The Rhythm Gimbal resembles a gyroscope. When it is still it is completely silent. When I pick it up and move it, it begins to emit sound. The sound is created by the rings rubbing against each other – they also change color at contact. Once set in motion, the Rhythm Gimbal will slow down, but will take a long time to stop completely. If I give the Rhythm Gimbal a good spin as I release it, it emits an extra set of noises which are more tinkly, and which slow down as the intrument winds down. Thus, unless I am careful to release it without any spin, it will continue to make sounds when I'm not looking at it. The 'background' sound heard while I am playing the other instruments comes from the Rhythm Gimbal.

Ich habe erst einen Monat vor der Aufführung mit dem Entwerfen der Virtuellen Welt von „The Sound of One Hand" begonnen, hatte auch nur soviel Zeit, sie spielen zu lernen, mußte mich also total in den Entstehungsprozeß versenken.

Eigentlich dachte ich, das Stück würde zu einer ausgefeilten VR (Virtual Reality)-„Demo", oder Erklärung, mit klar sichtbaren Einsätzen für die Musik, leicht zu handhabenden Interfaces und einem Haufen lustiger Rube Goldberg-Tricks. Aber als ich an der „Welt" arbeitete, kam eine Stimmung auf, irgendeine Essenz, was ziemlich mit meiner Gefühlslage oder spirituellen Erfahrung zu der Zeit zusammenhing. Das kam unerwartet und war aufregend, auch wenn es um nichts Fröhliches ging. Jedenfalls lief ein dunklerer, mehr intuitiver Prozeß in mir ab, anstatt mich der üblichen Computerkultur mit ihrer Geordnetheit und ihrem Witz einzufügen. Es hat nur wenige Augenblicke gegeben, in denen ich das Gefühl hatte, in einer intuitiven Weise zu programmieren. Das war einer davon. Erwarte von den Instrumenten nicht, daß sie unmittelbar verständlich oder einfach zu spielen sind. Sie entstammen einem schöpferischen Prozeß, den ich nicht völlig erklären kann und ich mußte erst lernen, sie zu spielen. Ich denke nicht, daß die beiden Ästhetiken, die ich unterscheide, sich gegenseitig ausschließen müssen. Aber der intuitive Anteil der Gleichung kann nicht per Knopfdruck erzwungen werden. Eine Synthese von Klarheit und Stimmung kommt, wenn überhaupt, aus der Eingebung.

Das erste Instrument heißt Rhythm Gimbal. Eine Gimbal ist eine gewöhnliche mechanische Konstruktion, eine Reihe rotierender Gelenke, sie ähnelt einem Kreisel. Wenn sie stillsteht, ist sie ganz weiß und völlig lautlos. Wenn ich sie aufhebe und bewege, fängt sie an, Töne von sich zu geben. Der Ton wird durch die aneinanderreibenden Ringe erzeugt; sie verändern auch ihre Farbe durch den Kontakt. Einmal in Bewegung, verlangsamt sich die Gimbal, aber es dauert lange, bis sie völlig stillsteht. Wenn ich der Rhythm Gimbal beim Loslassen einen ordentlichen Dreh gebe, macht sie noch weitere Geräusche, die eher klingeln und mit dem Ausdrehen des Instruments langsamer werden. Wenn ich es nicht vorsichtig absetze, so daß es keinen Drall bekommt, wird das Instrument, auch wenn ich mich etwas anderem zuwende, weiter Geräusche machen. Die „Hintergrundgeräusche", die man hört, während ich die anderen Instrumente spiele, kommen von der Rhythm Gimbal.

Der wichtigste (nicht-klingelnde) Ton der Rhythm Gimbal ist die Verbindung eines Chors, eines Orchesters und noch einiger anderer Dinge. Die Harmonie wird durch den Schwung erzeugt, bei dem die Innenteile des Instruments aneinanderschlagen, nachdem man es losgelassen hat. Jeder Ring überträgt den Schwung auf den nächsten äußeren Ring, was eine komplexe Bewegung schafft, ähnlich Pendeln, die an Pendeln aufgehängt sind. Auf den Ringen sitzen Perlen. Wenn die Perlen gegeneinander schlagen, wechseln sie die Farbe und rufen auch eine Veränderung der Harmonie hervor. Kennst Du diese alten Spiele im Vergnügungspark, wo man mit einem riesigen Holzhammer auf eine Zielscheibe am Boden haut und dann sieht, wie hoch man die Scheibe auf einer langen senkrechten Latte schicken kann? Die inneren Stöße der Rhythm Gimbal schleudern in ganz ähnlicher Weise virtuelle Scheiben um einen Quintenzirkel und dann die Naturtonreihe hinauf. Wenn die zwei Sorten Scheiben die Harmonie etwa gleichzeitig erreichen, wird ihr ein Ton hinzugefügt. Alle Harmonie und rhythmische Struktur erwachsen aus diesem Vorgang.

Aber die Gimbal kann nicht wirklich als algorithmische Musikerzeugung beschrieben werden. Zum Beispiel glaube ich nicht, daß die Gimbal so eindeutig initialisiert werden könnte, um die richtigen Parameter zu finden, die sie zum Klingen bringen. Es gibt ein notwendiges Element intuitiven Spiels in den schrägen Harmonien dieses komischen Instruments.

Jeder Ton des Stückes wird durch meine Handbewegungen erzeugt, die durch die virtuellen Instrumente übertragen werden: Es gibt keine vorbestimmten Sequenzen oder Tonfolgen; der musikalische Gehalt ist völlig improvisiert, mit Ausnahme des Klangumfangs der Instrumente.

Das heißt aber nicht, daß ich jede beliebige Musik machen kann. Das läuft genausowenig wie mit irgendeinem anderen Instrument. Ich kann keinen bestimmten Akkord per Knopfdruck aus der Rhythm Gimbal herausholen. Aber ich kann ein Gefühl aus einer Folge von Akkorden herausholen, weil ich beeinflussen kann, wann Akkorde wechseln und wie drastisch der Wechsel sein wird. Das bedeutet nicht weniger Kontrolle, sondern eine andere Art von Kontrolle. Der Prüfstein eines Instruments liegt nicht in dem, was es leistet, sondern darin, ob Du ihm gegenüber immer sensibler werden kannst, je mehr Du entdeckst und lernst. Ein Klavier ist genauso. Ein gutes Instrument hat eine Tiefe, die der Körper lernen kann, aber nicht der Verstand. Ich glaube, es ist dem Verstand völlig unmöglich, solche Instrumente zu erfinden.

In der VR sind versteckte Mechanismen lediglich unsichtbare Objekte. Während ich diese Welt entwickelte, machte ich die harmonischen Strukturen sichtbar. Das sieht aus wie eine Anzahl Noten, die auf Ringen herum- und einen Pfahl hinaufkrabbeln. Aber für die Aufführung machte ich sie im Rahmen des visuellen Designs größtenteils unsichtbar. Ein Teil ist immer noch sichtbar: ein großer

The primary (non-tinkly) Rhythm Gimbal sound is a combination of a choir, an orchestra and some other stuff. The harmony is generated by the momentum with which internal parts of the instrument hit each other after it has been released by the hand: each ring transmits spin to the ring outside it, creating a complex motion, like pendulums hung on pendulums. The rings have beads on them. When the beads collide, they change color, and also cause a change in harmony. You know those old attractions at amusement-park arcades, where you hammer a target on the ground with a giant mallet and see how high you can send a puck on a big vertical ruler? The internal collisions of the Rhythm Gimbal fling virtual pucks around the circle of fifths, and then up the harmonic series, in much the same way. A note is added to the harmony when the two types of puck reach it approximately at once. All the harmony and rhythmic texture come out of this process.

But the Gimbal can't properly be described as an algo-rythmic music generator. For example, I don't think an explicit style of initialization could be used to find the right parameters to make it sing. There is a necessary element of intuitive performance in the weird harmonies of this curious instrument.

Every note of the piece is generated by my hand movements, as they are transmitted through the virtual instrument: There are no predeterminded sequences or groupings of notes; the musical content is entirely improvised, with the exception of the timbral range of the instrument. This does not mean that I can make any arbitrary music, any more than I could with any other musical instruments. I can't get a specific chord out of the Rhythm Gimbal reliably. But I can get a feel out of a chord progression, because I can influence when chords change and how radical the change will be. This does not feel like less control to me, but rather like a different kind of control. The test of an instrument is not what it can do, but: can you become infinitely more sensitive to it as you explore and learn? A piano is like this. A good instrument has a depth that the body can learn and the mind cannot. I believe it is entirely impossible for the mind to invent such instruments.

Hidden mechanisms in Virtual Reality are just invisible objects. While I was developing this World, I would make the harmonie structure visible – it looks like a bunch of notes crawling on rings and up a pole. But I made it mostly invisible for the performance as a visual design decision. One part ist still visible, though: a lage blue ring with tuning forks on it. Each of the tuning forks has a T-shaped thing on the base and rings on the arms. These objects store the current legal tonic and chords for progressions; you can see them moving as the harmony changes.

The CyberXylo is a mallet instrument. Its notes are taken from the tuning forks on the blue ring, so it is always harmonious with the Rhythm Gimbal. The mallet retains angular momentum, with some friction, when it is released. Thus it is possible to set it spinning so that it will continue to hit the keys of the CyberXylo on its own for a while. The spin is of poor mathematical quality: it increments rotations instead of using quaternians. This creates wild, unnatural spinning patterns. With practice, enthusiastic spins of the mallet close to the keys can be a source of remarkable rhythms.

The Cybersax is the most ergonomically complex instrument. When the instrument is grabbed, it turns to gradually become held correctly by your hand and tries to avoid passing through fingers on the way. Once you are holding it, the positions of your virtual fingers continue to respond to your physical ones, but are adjusted to be properly placed on the sax keys. This is an example of a "simulation of control" that is critical in the design of virtual hand tools, especially when force-feedback is not available.

Three musical registers – soprano, alto, and bass – are located along the main tube. Each register consists of a set of shiny sax-like keys. The notes played by the keys come from the current set of legal notes defined by the Rhythm Gimbal, so it will not clash with the other instruments. It is possible to slide between registers by jerking the hand toward a targeted register. The momentum of your slide helps determine which notes will be associated with the keys until you slide again (if, for example, you approach the soprano register from the alto with greater force you will choose a set of notes that are higher up in pitch). You can play freely without dropping the horn by mistake (this was a hard quality to achieve). In the upper register, it is possible to play two melodies at once by modulating with the thumb. The orientation of the horn in space controls the timbre, mix, and other properties of the sound. Other design elements include the obscene, wagging tail/mouthpiece and the throbbing bell.

The Cybersax sound and geometric construction were partly inspired by a bizarre bamboo saxophone I have that came from Thailand. It is jointed at the top, just like the Cybersax's tail. Computer music must use instruments built out of concepts

blauer Ring mit Stimmgabeln drauf. Jede der Stimmgabeln hat ein Ding in Form eines T an seiner Basis und Ringe an den Armen. Diese Objekte speichern die jeweils geltende Toniken und die Akkordfolgen; man kann sehen, wie sie sich mit dem Wechsel der Harmonie bewegen.

Das CyberXylo ist ein Schlaginstrument. Seine Töne werden von den Stimmgabeln auf dem blauen Ring abkopiert, deshalb steht es immer im Gleichklang mit der Rhythm Gimbal. Dem Hammer bleibt, nachdem er losgelassen wird, einiger Schwung mit ziemlich viel Reibung. Es ist dadurch möglich, ihm einen Impuls zu geben, so daß er selbst noch eine Weile die Tasten des CyberXylos schlägt. Der Schwung hat kaum mathematische Qualitäten aufzuweisen. Er fügt Umdrehungen hinzu, anstatt Quaternionen zu gebrauchen. Das erzeugt wilde, chaotische Rotationsmuster. Bei einiger Übung können, nahe an den Tasten, begeisterte Wendungen des Hammers zur Quelle erstaunlicher Rhythmen werden.

Das Cybersax ist das ergonomisch komplizierteste Instrument. Wenn man nach dem Instrument greift, dreht es sich so, daß die Hand es nur richtig anfassen kann, und verhindert dabei, daß es Dir durch die Finger gleitet. Wenn Du es einmal hältst, werden die Positionen Deiner virtuellen Finger weiter auf Deine physischen Finger reagieren, sind aber darauf ausgerichtet, sich korrekt auf die Tasten des Saxophons zu legen. Dies ist ein Beispiel einer „Simulation von Kontrolle", die im Design einfacher virtueller Werkzeuge entscheidend ist, besonders wenn kein Kraft-Feedback vorhanden ist.

Drei Stimmlagen – Sopran, Alt und Bass – sind entlang des Hauptrohrs angelegt. Jede Stimmlage besteht aus einem Satz glänzender saxophonähnlicher Tasten. Die von den Tasten gespielten Laute entstammen dem jeweils anerkannten Satz Töne, der durch die Rhythm Gimbal definiert wird. Dadurch harmonieren sie mit den anderen Instrumenten. Es ist möglich, zwischen den Stimmlagen hin und herzugleiten, indem man mit der Hand auf die angepeilte Stimmlage langt. Der Schwung Deiner Bewegung bestimmt mit, welche Taste mit welchen Tönen belegt wird – bis Du wieder woandershin gleitest. (Wenn Du Dich zum Beispiel vom Alt dem Sopran ziemlich heftig näherst, wirst Du einen Satz Töne mit höherer Stimmung wählen). Du kannst frei spielen ohne das Instrument aus Versehen fallenzulassen. (Diese Eigenschaft war ziemlich schwer hinzukriegen). In der oberen Stimmlage ist es möglich, durch das Modulieren mit dem Daumen zwei Melodien gleichzeitig zu spielen. Die räumliche Ausrichtung des Sax bestimmt Farbe, Mischung und andere Eigenschaften des Klangs. Weitere Gestaltungselemente sind u.a. das obszöne schwanzwedelnde Mundstück und die erbebende Glocke.

Sound und geometrischer Aufbau des Cybersax wurden zum Teil durch ein bizarres Bambussaxophon, das ich aus Thailand habe, inspiriert. Es ist oben verbunden, genauso wie das Endstück des Cybersax.

Computermusik muß Instrumente benutzen, die aus Konzepten darüber, was Musik ist, konstruiert sind. Das ist eine drastische Abwendung von den „dummen" Instrumenten der Vergangenheit. Ein Klavier weiß nicht, was eine Note ist, es vibriert eben, wenn es angeschlagen wird. Eine Sensibilität und eine Achtung für das Geheimnis des Lebens sitzen im Herzen der Wissenschaft und auch der Kunst. Instrumente mit eingebauten, fixen Vorstellungen können diese Sensiblität abstumpfen, indem sie der Handlung ein scheinbar plattes Setting geben. Das kann zu „blöder" oder fader Kunst führen. Es ist interessant, sich hinter einem Klavier zu verstecken statt hinter einem Computer, aber nur, weil ein Klavier aus tönenden Materialien gemacht ist, nicht aus Konzepten. Damit Computerkunst oder -musik funktionieren, mußt Du besonders bemüht sein, Leute und menschliche Bezüge in den Mittelpunkt des Interesses zu stellen.

Ich war total froh, als ich merkte, daß „The Sound of One Hand" eine ungewöhnliche Beziehung zwischen der jeweiligen Rolle von Performer, Publikum und Technologie schuf. Normalerweise wird seltene und teure Hochtechnologie in der Performance gebraucht, um ein Spektakel zu erzeugen, das den Status des Performers erhöht. Er wird ziemlich unerreichbar gemacht, während das Publikum baß erstaunt zu sein hat. Darin gleichen sich Rockkonzerte und der Golfkrieg.

„The Sound of One Hand" erzeugt eine ziemlich andere Situation. Das Publikum sieht mit an, wie ich mich verrenke, während ich den Raum durchschiffe und die virtuellen Instrumente bediene. Ich aber trage einen Datenhelm (Eye-Phones). Fünftausend Menschen beobachten mich, aber ich kann sie nicht sehen und weiß auch nicht, wie sie mich finden. Ich war verletzlich, trotz der Technologie. Das erzeugte einen authentischeren Kontext für die Musik. Wenn Du je Musik vor einem Publikum gespielt hast, besonders improvisierte Musik, kennst Du die Art von Verletzlichkeit, die ich meine, die einer authentischen Aufführung vorausgeht.

Zu diesem Verrenken ... Ich habe in der Performance Point-Flying benutzt. Das ist eine Navigationstechnik, bei der Du mit Deiner Hand auf einen Punkt deutest, an den Du gelangen möchtest. Das bringt Dich dazu, dorthin zu fliegen. Ich kann Point-Flying in den industriellen Anwendungen der VR nicht leiden – es erfordert Geschicklichkeit und Du hast Deine Hand nicht frei für andere Dinge. Hier habe ich es benutzt, weil ich genau diese ungebundene, geschickte Navigationsweise wollte. Dadurch konnte ich, parallel zur Performance, eine Reise durch die Asteroiden der virtuellen Welt choreographieren. Es war mir irre peinlich und ich war ziemlich geschockt, als ich mich bei einer der Performances in meiner eigenen Welt verlief!

Ein weiteres menschliches Element des Stückes ist seine Körperlichkeit. „The Sound of One Hand" steht in der Tradition des Theremion, da das Interface hauptsächlich physisch statt mental ist. Obwohl die Instrumente aus

of what music is. This is a drastic departure from the "dumb" instruments of the past. A piano doesn't know what a note is, it just vibrates when struck. A sensitivity to, and a sense of awe at, the mystery that surrounds life is at the heart of both science and art; instruments with mandatory concepts built in can dull this sensitivity by providing an apparently non-mysterious set ting for activity. This can lead to "nerdy" or vland art. It is interesting to hide one's self behind a piano, as opposed to a computer, but only because a piano is made of resonant materials, not of concepts. In order for computer art, or music, to work you have to be extra-careful to put people and human contact at the center of attention.

I was delighted to discover that The Sound of One Hand created an unusual status relationship between the performer, the audience, and the technology. The usual use of rare and expensive high technology in performance is to create a spectacle that elevates the status of the performer. The performer is made relatively invulnerable, while the audience is supposed to be awestruck. This is what rock concerts and the Persion Gulf War have in common. The Sound of One Hand creates quite a different situation. The audience watches me contort myself as I navigate the space and handle the virtual instruments, but I am wearing Eye-Phones. Five thousand people watch me, but I can't see them, or know what I look like to them. I was vulnerable, despite the technology. This created a more authentic setting for music. If you have played music, especially improvised music, in front of an audience, you know the kind of vulnerability I am talking about, the vulnerability that precedes an authentic performance.

About that contorting ... I used point-flying in the performance. This is a technique of navigating where you point with your hand to where you want to go and this causes you to fly there. I dislike point-flying in industrial applications of VR – it requires skill and uses up your hand. I used it in this case because I did want the unconstrained, skillful type of navigation; it allowed me to choreograph a tour of the Virtual World's asteroid along with the performance. I was shocked and embarrassed when I got lost in my own World during one of the performances! Another human element of the piece is its physicality. The Sound of One Hand is in the tradition of the theremin in that the interface is primarily physical instead of mental. Although the instruments were made of information, the music was primarily made of gesture.

The equipment I used was, for the most part, not state-of-the-art, but about a year out of date. The synthesizers and Head Mounted Display were '92 models, but the graphic engine, tracker and DataGlove were all older. I think you have to actively avoid using the latest gear in doing art, to avoid getting caught up in technology for its own sake.

The software was quite current, however. The piece was written entirely in Body Electric, a visual programming language for Virtual Reality. I am extremely fond of this software working environment, which was designed primarily by Chuck Blanchard. You hook up visual diagrams to control what happens in the Virtual World and see the effect immediately. All the music and physics were done in Body Electric; I could never have made this thing in C.

Visually and sculpturally, the World took advantage of every trick then available for real-time rendering, including radiosity, fog, texture mapping, environment mapping, and morphing. The color "flaking" effect resulted from a bug seen when the color of hardware fog on the graphics engine was gradually changed (I set up a very slow-moving bouncing ball in the cylinder of red/green/blue color space as a chooser for the fog color). I sculpted all the parts of the World except for the illuminated skeletal hand that sprouts from the asteroid wall – which is from a Magnetic Resonance scan of a patient's hand taken at the Veterans Administration hospital in Palo Alto. It was originally used in research on surgical simulation. The asteroid is hollow, and about twelve feet in diameter, although the dense fog makes it look and feel immense. It has a big crack in the side, through which fireflies frolic, a big red ginger plant growing inside, and also a few spotlights. The instruments are generally kept inside. At one point in the performance I spun the Gimbal and flew out through the crack for a while to let the audience see how lonely the asteroid was, surrounded by absolute void.

The sounds were created on two sampler/synthesizers. I decided not to use the wonderful 3D sound capabilities of the Virtual Reality, since they are intended primarily for headphone use, and I didn't want the audience to be trying to hear something.

In many ways, The Sound of One Hand was a bigger leap into the unknown than all of the weird "experimental" performances I had been involved in in New York in the late seventies. I had no idea if the piece would take on a mood of

Information gemacht wurden, bestand die Musik aus Gebärden.

Fast alle Geräte, die ich benutzt habe, waren nicht mehr am neuesten Stand der Technik, sondern etwa um ein Jahr veraltet. Die Synthesizer und das Head Mounted Display waren 92er Modelle, doch die Graphik-Engine, der Tracker und der Datenhandschuh waren alle älter. Ich glaube, in der Kunst sollte man der Verwendung der neuesten Geräte bewußt aus dem Weg gehen, um zu verhindern, daß man der Technologie um ihrer selbst willen verfällt.

Die Software aber war ziemlich neu. Das Stück war komplett in Body Electric geschrieben, einer visuellen Programmiersprache für Virtual Reality. Ich mag die Arbeitsumgebung dieser Software, die hauptsächlich von Chuck Blanchard entwickelt wurde, besonders gern. Du schließt visuelle Diagramme zusammen, um zu kontrollieren, was in der virtuellen Welt passiert, und Du siehst den Effekt sofort. Die gesamte Musik und die Physik sind in Body Electric geschrieben; ich hätte dieses Ding nie auf C machen können.

Die Welt zog, visuell und plastisch, Nutzen aus jedem damals für Darstellungen in Echtzeit zur Verfügung stehenden Trick, einschließlich Radiosität, Nebel, Mapping von Textur und Umwelt, und Morphing. Die Farbflockenbildung ergab sich aus einem Programmierfehler, der sichtbar wurde, als die Farbe des Hardwarenebels auf der Graphik-Engine langsam verändert wurde. (Ich setzte einen sich sehr langsam bewegenden, hüpfenden Ball in den Zylinder des rot/grün/blauen Farbraumes als Chooser für die Nebelfarbe). Ich habe alle Teile der Welt gebildet, bis auf das beleuchtete Handskelett, das aus der asteroidischen Wand wächst. Es stammt von einem MR-Scan der Hand eines Patienten aus dem Veterans Administration – Krankenhaus von Palo Alto. Es war ursprünglich in der Forschung zur Operationssimulation verwendet worden.

Der Asteroid ist hohl und hat einen Durchmesser von etwa 4 Metern, obwohl der dichte Nebel ihn riesig erscheinen läßt. Durch einen großen Riß an der Seite schwirren Glühwürmchen. In seinem Inneren wächst eine große rote Ingwerpflanze; es gibt auch ein paar Scheinwerfer. Die Instrumente werden meistens drinnen aufbewahrt. An einem Punkt der Performance gab ich der Gimbal einen Schwung und flog für eine Weile durch den Riß ins Freie, damit das Publikum sehen konnte, wie einsam der Asteroid war, von absoluter Leere umgeben.

Die Klänge wurden auf zwei Abtastern/Synthesizern erzeugt. Ich beschloß, die wunderbaren dreidimensionalen Klangmöglichkeiten der VR nicht zu verwenden, da sie vor allem für Kopfhörergebrauch entwickelt worden sind. Ich wollte nicht, daß das Publikum versucht, etwas zu hören.

In vieler Hinsicht war „The Sound of One Hand" ein größerer Sprung ins kalte Wasser als all die verrückten „experimentellen" Performances, an denen ich Ende der siebziger Jahre in New York beteiligt war. Ich wußte vorher nicht, ob das Stück eine Stimmung oder eine Bedeutung annehmen würde, oder ob das Publikum in der Lage wäre, die Erfahrung zu verarbeiten. Für mich wurde die Performance zu einem fröhlichen, therapeutischen Ereignis. Es war so eine Art technologischer Blues, ein trostloses Werk, das ich glückselig spielen konnte. Es war eine Gelegenheit, mit der VPL-Familie an einem rein kreativen Projekt zu arbeiten, eine Möglichkeit, den ganzen VPL-Kram als gegebenes (verläßliches) Rohmaterial zu sehen statt als zu erledigende Arbeit. Es war eine Chance, mein Theoretisieren über virtuelles Werkzeugdesign in die Praxis umzusetzen, VR nur um seiner Schönheit willen anzuwenden, und gegenüber meiner lächerlich politischen Kollegenschaft einfach nur Musik zu machen. Es war auch eine Feier, weil ich VPL nicht mehr leiten mußte. Das Publikum war immer unglaublich aufgeschlossen und niemand hat, soweit ich weiß, das Stück als Demo beschrieben. Es wurde als Musik erlebt.

meaning or if the audience would find the experience comprehensible. The performance turned out to be a cheerful, therapeutic event for me. It was a sort of technological blues, a bleak work that I could play happily. It was a chance to work on a purely creative project with the VPL family, a chance to treat all of VPL's stuff as a given set of (reliable!) raw materials instead of as work to do, a chance to practice what I preach about virtual tool design, a chance to use VR just for beauty, and a chance to be musical in front of my ridiculously political professional peer community. It was also a celebration of not having to run VPL anymore. The audiences were incredibly responsive, and I didn't hear anyone describe the piece as a demo. It was experienced as music.

Welt und Musik: Jaron Lanier
System und Unterstützung bei der Aufführung: Dale McGrew
Synthesizer und Tontechnik: Alfred „Shabda" Owens
Body Electric (Weltdesign und Steuerung): Chuck Blanchard, David Levitt
Isaac (Echtzeit Graphik Software): Ethan Jaffe, Chris Paulicka
World Test Intern (Gimbal Kreisel): Rolf Randa
Datenhelm, Datenhandschuh, Entwicklungssoftware:
VPL Research, Inc.
Graphik Engine (440VGXT): Silicon Graphics, Inc.
Wir danken Joanneum Research für die freundliche Unterstützung des Projekts.

World and Music: Jaron Lanier
System and Show Support: Dale McGrew
Synth. and Sound Engineer: Alfred "Shabda" Owens
Body Electric (world design and control tool):
Chuck Blanchard, David Levitt
Isaac (real-time graphics software): Ethan Joffe, Chris Paulicka
World Test intern (Gimbal Spinner): Rolf Rando
EyePhones, DataGlove, development software:
VPL Research, Inc.
Graphics Engine (440VGXT): Silicon Graphics, Inc.

X-TOPIA

ELLIOTT SHARP
Soldier String Quartet

Extopia is *The Thing Itself*, downloaded and transformed. 'Outside of its place' – it exists as extopia. This is not an endproduct or a static process or a one-shot event, but a feedback loop.

Transformation was a key element in *CRYPTID FRAGMENTS* (found on my eponymous CD on the Extreme label, XCD 020.) In that piece, core materials were played by violin and cello, dumped into the computer and radically sculpted through time-expansion compression, pithshifting, filtering, reversing, editing, and recombining. Some of the resultant 'instruments' retained much of their identity, some were rendered unrecognizable as string sounds (here 'instrument' will be redifined to mean the final sound or phrase created by processing). All of the instruments were finally ordered in time through layering and playlists to yield a virtual string-quartet. This quartet could not, however, be performed in real-time. All of its transformations are processor-intensive and require rendering and editing time.

I had long sought to perform such a transformative electroacoustic piece in real-time – in fact, much of my performance on guitar (beginning in 1969) used electronic and mechanical processing to place the sounds 'outside' of their normal form and role as guitar. With the advent of the PC, my work in this zone (under the name *VIRTUAL STANCE* – 1986 – 1990) entered the digital realm and made use of the software M running on an Atari 1040 to drive various samplers and digital delays. With M, an improvisational environment could be prepared and then realized in performance. M was a clocked sequencer – there was a tendency to rhythmic structures involving cyclic repetitions. I tried to make these tendencies less obvious through the use of samples of my invented string instruments (slab, pantar, violinoid) that had been processed to mask their origins.

X-TOPIA further refines this real-time approach. The string quartet is given core materials ordered in time as well as an instruction set of musi-

Extopia ist *Das Ding An Sich*, heruntergeladen und transformiert. „Außerhalb seines Ortes" – es existiert als Extopia. Es ist kein Endprodukt, kein statischer Prozeß, kein einmaliges Ereignis, sondern eine Feedback-Schleife.

Transformation ist ein Schlüsselelement in *CRYPTID FRAGMENTS* (enthalten auf meiner auf dem Label Extreme erschienenen eponymen CD XCD 020). Das Kernmaterial des Stücks wurde auf Geige und Cello gespielt, in den Computer eingegeben und dann durch Zeitverzögerung und -kompression, Verschiebung der Tonhöhen, Filtern, Rückwärtsspielen, Schneiden und erneutes Kopieren radikal bearbeitet. Einige der „Instrumente", die so entstanden, behielten viel von ihrer ursprünglichen Identität, andere dagegen waren nicht mehr als Streicherklänge erkennbar (der Begriff „Instrument" ist hier definiert als der Klang oder die Phrase, die durch die Bearbeitung entstanden ist). Alle Instrumente wurden dann durch Schichtung und Playlists zeitlich geordnet zu einem virtuellen Streichquartett. Dieses Quartett konnte allerdings nicht in Echtzeit aufgeführt werden. Die Transformationen sind prozessorintensiv und brauchen Speicherkapazität für Wiedergabe und Schnitt.

Ich wollte schon lange solch ein transformatives elektroakustisches Stück in Echtzeit aufführen – und bei dem, was ich (seit 1969) mit der Gitarre machte, verwendete ich immer wieder elektronische und mechanische Nachbearbeitungstechniken, um die Klänge „außerhalb" ihrer normalen Form und Rolle zu plazieren. Mit dem Aufkommen der PCs verschob sich meine Arbeit (unter dem Namen *VIRTUAL STANCE* – 1969 – 1990) in den Digitalbereich, und über die Software M auf einem Atari 1040 steuerte ich verschiedene Samplers und digitale Delays. Mit M konnte ich eine Improvisationslandschaft vorbereiten und diese dann auf der Bühne realisieren. M war ein Clocked Sequenzer – und bedingte rhythmische Strukturen zyklischer Repetitionen. Mit Samples meiner erfundenen Streichinstrumente (Slab, Pantar, Violinoid), deren Ursprung verschleiert worden war, versuchte ich, diese Strukturen weniger auffällig zu machen.

In *X-TOPIA* ist dieses Echtzeit-Konzept noch weiter entwickelt. Das Streichquartett bekommt zeitlich geordnetes Kernmaterial sowie Anweisungen für musikalische Aktionen. Durch den Buchla Thunder (mit seinen hervorragenden Echtzeit-MIDI-Control-Möglichkeiten) als Controller und verschiedene Digitalsignalprozessoren können die Klänge des Quartetts gesampelt und vom Komponisten auf der Bühne klanglich und räumlich transfiguriert werden. Das Quartett entscheidet auf Grund der durch meine Bearbeitung entstandenen Klangprozesse über seine weiteren musikalischen Aktionen im Stück, die dann ihrerseits bearbeitet und in den klingenden Raum geschickt werden.

Die Wortkreation „Ir/rationale Musik" beschreibt treffend, was ich mit meinen Kompositionen ausdrücken möchte:

Ir – die Klangakustik im Raum und im Ohr (in space and in the ear) und ihre Verbindung zur Wahrnehmungsmaschine: Obertonreihe, Differenztöne, Feedback, Lautstärkeneffekte, „define melody", „define groove";

das Rationale: Struktur und Ordnung, Verwendungs- und Bearbeitungsalgorithmen, formale Organisationssysteme, gesellschaftlicher und gattungsspezifischer Kontext, Querverweise.

Insgesamt – das Ir/rationale – Chaos, Intuition, das Tangentiale.

Nicht-lineare Improvisation ist ein wesentlicher Teil ir/rationaler Musik – Improvisation erweckt die Musik zum Leben; sie macht Statisches dynamisch; der individuelle Gestalter im Klangfluß.

cal actions. Using the Buchla Thunder (offering unprecedented real-time MIDI control) as a controller and various digital signal processors, the sounds of the quartet may be sampled and transfigured sonically and spatially in performance by the composer. The sonic processes instigated by my processing are integrated by the quartet to effect their own decisions about further musical actions to be taken within the piece, which are again processed and sent into the sonic field.

Irrational music is a useful pun to describe what I try to manifest in my compositions: ir (an irresistible pun) – the acoustics of sound in space and in the eye and its connection to the perceptual engine: the overtone series, difference tones, feedback, volumn effects, 'define melody', 'define groove'.

the rational: structure and order, algorithms of use and process, formal systems of organization, social and genre context, cross-reference.

overall – the irrational – chaos, intuition, the tangential.

Non-linear improvisation is an essential part of irrational music – improvisation brings the music to life; it transforms the static into the dynamic; the individual effector in the sonic flux.

Oily Sam

KEN VALITSKY
Soldier String Quartet

Oily Sam was commissioned by the Soldier String Quartet in the Autumn of 1993. The piece was written for string quartet and interactive computer software, and utilizes reactions of the players to sounds emanating from the computer. Sections of the composition employ improvisation in what are idiomatically American musical genres. Jazz and rock elements are employed, both in contrast to and agreement with the music generated by the computer. In addition, quotations taken from a variety of sources, i.e. Mozart, Elliot Carter and others, are used as generative material. These quotations are rhythmically and texturally altered throughout the piece. They are also used as germatic cells given to the performers in order to create larger improvisatory sections.

Throughout the composition, different vocal quotations (samples) are utilized in order to underline the basic premise of *Oily Sam*. Politics and politicians are laughable yet intrinsically dangerous. Samples of Bill Clinton, Ronald Reagan, Richard Nixon and Rocky and Bullwinkle illustrate the basic absurdity of politics.

Through juxtaposition of political quotes with comments from cartoon characters, the complete buffoonery of politicians, their statements, and their long-term accomplishments are exposed. Although politics is ridiculed throughout the piece, the listener is reminded at the end of the absolute evil which is possible with the acquisition of power and everything it entails.

Musically, *Oily Sam* is a continuation of a style which I have been developing over the last five years. Elements of a variety of mainly American musical genres such as heavy metal, be-bop, rap, hip-hop and blues are used as formative material. Also, world music such as Northern Indian Classical music and different styles of Sub-Saharan African music are employed to show the basic relationships shared by all music. Structurally, *Oily Sam*'s form was determined through the use of the Golden Section. The number .618 was used, not as composers such as

Oily Sam entstand im Auftrag des Soldier-String-Quartet im Herbst 1993. Es wurde für Streichquartett und interaktive Computersoftware geschrieben und verwertet die Reaktionen der Musiker auf Klänge aus dem Computer. Zum Teil basiert die Komposition auf Improvisationen innerhalb idiomatischer amerikanischer Musikgenres. Verwendet werden Jazz- und Rockelemente, sowohl kontrastiv zur computergenerierten Musik als auch in harmonischer Korrelation. Als generatives Material werden außerdem Zitate aus verschiedensten Quellen wie Mozart, Elliot Carter u.a. herangezogen. Diese Zitate werden im Verlauf des Stückes rhythmisch und strukturell verändert und den Musikern als Keimzellen zur Entwicklung längerer Improvisationspassagen vorgegeben.

In die Komposition integrierte gesprochene Zitate (Samples) unterstreichen das Grundprinzip von *Oily Sam*. Die Politik und die Politiker sind lächerlich, zugleich aber inhärent gefährlich. Samples von Bill Clinton, Ronald Reagan, Richard Nixon sowie Rocky und Bullwinkle illustrieren die grundlegende Absurdität der Politik. Durch die Nebeneinanderstellung von politischen Aussagen und Sprüchen von Cartoon-Figuren wird die Komik der Politiker, ihrer Statements und ihrer längerfristigen Leistungen offenbar. Obwohl die Politik im gesamten Stück ins Lächerliche gezogen wird, wird dem Zuhörer am Ende doch bewußt, wie durch die Erlangung von Macht und allem, was damit zusammenhängt, das absolute Böse möglich wird.

Musikalisch gesehen ist *Oily Sam* die Fortsetzung eines Stils, den ich in den letzten fünf Jahren entwickelt habe. Als formatives Material habe ich Elemente einer Reihe hauptsächlich amerikanischer Musikgenres verwendet, wie z.B. Heavy Metal, Be-Bop, Rap, Hip-Hop und Blues. Weltmusik-Richtungen wie nordindische klassische Musik oder verschiedene Stilrichtungen aus afrikanischen Regionen südlich der Sahara dokumentieren die jeder Art von Musik gemeinsamen Beziehungen. Strukturell ist die Form von *Oily Sam* durch die Anwendung des Goldenen Schnitts determiniert. Die Zahl 0,618 dient nicht wie bei Bartók oder Debussy zur Bestimmung der Abschnittslänge durch die Taktanzahl, sondern als Echtzeit-Controller von Proportionen. Auf makroarchitektonischer Ebene besteht *Oily Sam* aus zwei Abschnitten. Jeder Abschnitt besteht aus Unterabschnitten, die ihrerseits wieder unterteilt sind, und zwar solange, bis auf diese Art das kleinste musikalische Element bestimmt worden ist. So steht jeder Abschnitt und Unterabschnitt strukturell zu den Gesamtproportionen des ganzen Stückes in Relation (siehe Diagramm).

Außer dem Streichquartett wurde für das Stück das folgende Equipment verwendet: ein Macintosh IIci zur Steuerung von Sample Cell, ein Ensoniq EPS 16 plus, ein Yamaha TX802, ein Oberheim Matrix 1000, ein E-Mu Pro/Cussion und ein Kawai K1R. Kein einziger Ton, auch nicht von den Samples, blieb unverändert. Jedem Ton wurde etwas hinzugefügt, entweder durch Nachbearbeitung, Loops mit verschiedenen Crossfades oder andere digitale Manipulationsmittel.

Oily Sam ist eine gleichzeitig pessimistische und optimistische Sicht der Politik und der Folgen des Machtmißbrauchs. Pessimistisch, weil wir daran erinnert werden, daß menschliches Leid fast immer durch Politiker und ihre Politik verursacht wird. Optimistisch, weil wir trotzdem über sie lachen können.

Bartok and Debussy have used it to determine section length through number of measures, but rather as a real-time controller of proportion. *Oily Sam*, on a macro-architechtonic level is divided into two sections. Each section is in itself subdivided a number of times. This process continues within each subdivision until, finally, the smallest musical moment has been determined. In this manner each section and subsection relates structurally to the overall proportions of the entire piece (see diagram).

Technologically, the equipment used to create this piece, other than the string quartet, was a Macintosh IIci computer which controlled Sample Cell, an Ensoniq EPS 16 plus, a Yamaha TX802, an Oberheim Matrix 1000, an E-Mu Pro/Cussion and a Kawai K1R. Not one sound, including the samples, was left unaltered. Either through processing, looping with a variety of crossfades or through other means of digital manipulation, each sound has had something added to it.

Oily Sam is simultaneously a pessimistic/optimistic view of politics and the consequences of misuse of power. Pessimistic in that it is a reminder that nearly all of mankind's suffering has been a result of politicians and their policies. Optimistic in that we can still laugh at them.

KATHARSIS
catharsis

Computer-Konzert in sinfonischer Größenordnung
computer music concert of symphonic magnitude

GÜNTHER RABL

AUSTRIAN SOUNDSCAPE
Martin / Kaufmann / Rabl
Lichtgestaltung: Rainer Jessl
Raumgestaltung: Offenes Kulturhaus,
Gerhard Neulinger

A recording of a waterfall forms the basis of the full-length composition; however, more importantly than its function as acoustic material, it serves as an inexhaustible reservoir of forms and processes which can be derived from it by means of various analytical methods.

This recording is complemented by a number of "imaginary acoustic objects" – numerical oscillators (so-called physical models) designed especially for this purpose which are brought to life by the actual sound of the water and its transformations (often merely tiny excerpts).

In principle, this music is oriented to space: It was conceived for performance in a real space. All parts utilize 8 independent sound tracks which can be assigned to other groups of loudspeakers in any section of the spatial production.
Level 1: "KATHARSIS" 13 min.
Level 2: "FUNKENFLUG" 19 min.
Level 3: "WIND" (in preparation) 6 min..
Level 4 (the last): "GROSSE FUGE" (in preparation) 30 min.

For years, I have worked again and again with a single piece of material: the recording of a thundering waterfall in the mountains of the Salzburg province. The four pieces named "Toccata" from

Die Aufnahme eines Wasserfalles bildet die Grundlage für die gesamte, abendfüllende Komposition: aber nicht nur als Klangmaterial, sondern vor allem auch als unerschöpfliches Reservoir an Formen und Verläufen, die mit verschiedenen Analyseverfahren daraus abgeleitet werden können.

Demgegenüber steht eine Anzahl 'imaginärer Kunstprojekte' – eigens dafür gestaltete numerische Oszillatoren (sogenannte physikalische Modelle), die vom Klang des Wassers und dessen Transformationen (oft nur winzige Ausschnitte) zum Leben erweckt werden.

Diese Musik ist grundsätzlich raumbezogen: für die Aufführung in einem realen Raum konzipiert. Alle Teile verwenden 8 unabhängige Tonspuren, die aber im Rahmen der räumlichen Inszenierung in jedem Abschnitt anderen Gruppen von Lautsprechern zugeordnet werden können:
1. Stufe: „KATHARSIS" 13 min
2. Stufe: „FUNKENFLUG" 19 min
3. Stufe: „WIND" (in Arbeit) 6 min
4. und letzte Stufe: „GROSSE FUGE" (in Vorbereitung) 30 min

Seit Jahren arbeite ich immer wieder mit einem Material: der Aufnahme des Tosens eines Wasserfalles in den Salzburger Bergen. Die vier 'Toccata' benannten Stücke aus 'FAREWELL TEMPERED PIANO' sind daraus gewonnen, 'Aufstieg' und 'Abstieg' in 'ODYSSEE', und vieles andere mehr, sowohl im Bereich des Klanges als auch im

Bereich von Formen und Prozessen. In 'KATHARSIS' ist dieses Rauschen das einzige externe Klangmaterial, der 'Rohstoff', aus dem fast alles gewonnen wird. Klänge, Klangprozesse, Melodien, Rhythmen, räumliche Aufteilungen, etc.

Mit dynamischen Oszillatoren lassen sich Klangobjekte definieren, die im Medium sind, deren Klangwelt tatsächlich zur Gänze verfügbar ist, sogenannte 'imaginäre Objekte'. Sie sind zunächst stumm, bieten nur ein ganz bestimmtes Kontinuum von Schwingungsmöglichkeiten. Erst die Art der Initiierung entscheidet über das tatsächliche Resultat.

Das deutlichste Beispiel ist 'FUNKENFLUG', der zweite Teil von KATHARSIS. Hier habe ich neun Objekte definiert (nach dem Modell einer präparierten Saite), jedes sorgfältig festgelegt in Größe, Spannung, Elastizität, Ausklingverhalten und vielem mehr – ein Ambiente von aufeinander abgestimmten 'Gegenständen', die sofort als solche erkennbar sind. Durch die 'Funken' werden sie auf die verschiedenste Weise zum Klingen gebracht, denn diese Funken haben verschiedene Gestalt, können schwer oder leicht sein, langsam oder schnell, können an verschiedenen Stellen aufschlagen, können das Objekt in Ruhe treffen oder während es noch vom vorigen Schlag in Schwingung ist. Vor unserem geistigen Auge entsteht ein imaginäres Klangobjekt.

"FAREWELL TEMPERED PIANO" were composed on that basis, as well as "Aufstieg" ("Ascent") and "Abstieg" ("Descent") in "ODYSSEE" ("ODYSSEY") and many others, both in the area of sound and in that of forms and processes. In "KATHARSIS", this roar is the only external acoustic material, the "raw material" from which almost everything is produced: sounds, sound processes, melodies, rhythms, spatial distributions, etc.

ODYSSEE
Günther Rabl
Lichtgestaltung: Rainer Jessl
Raumgestaltung: Offenes Kulturhaus, Gerhard Neulinger

Fotos © Renate Porstendorfer

With Dynamic Oscillators, one can define acoustic objects in the medium, the acoustic world of which is completely available, so-called "imaginary objects". They are silent at first, offering only a certain continuum of possible vibrations. The type of initiation is what determines the actual result.

The most obvious example is "FUNKENFLUG", the second part of "KATHARSIS". In this section, I defined nine objects (according to the model of a prepared string), and each of them were recorded carefully with regard to size, tension, elasticity, sustaining qualities, and much more, thereby creating an ambience of harmonized "objects" which can be immediately recognized as such. They are sounded in many different ways by the "sparks", since these sparks have various forms; they can be light or heavy, fast or slow. They can strike different spots; they can strike the object when it is at rest or while it is vibrating from the previous stroke. An imaginary acoustic object is created before our spiritual eyes.

Komposition 1991 – 94, Dauer ca. 70 min
Klangprojektion über 8, 20 oder mehr unabhängige Kanäle

Composition: 1991 – 94; Duration: approx. 70 min.
Sound projection: over 8, 20 or more independent channels

ELECTRO CLIPS

CHRISTIAN MÖLLER / STEPHEN GALLOWAY

Description of the environment

"Electro Clips" is an installation for ballet which enables the dancers to interact with light and sound directly.

The visual and acoustic symbols which ballet normally uses as givens are produced and influenced in "Electro Clips" by the dancer Stephen Galloway, his movements and choreography.

A parallel environment of sound, light and movement will be produced. The dancer assumes the role of a director in that he can use the changing functions of the sensors distributed about the stage area like a keyboard in manipulating the sound and light.

Electronic control

Photoresistors set into the stage floor put out a variable resistance value depending on the actual amount of light shining on the respective part of the stage. Measurements are made of the resulting, variable compositions of light and shadow created by the dancer's movements when he enters the cone of light, which are visible to the audience. The respective amount of light remaining is transferred to the computer system in the form of electrical voltage, digitalized and then transmitted to the audio system as MIDI information.

Lighting

"Electro Clips" utilizes various types of light sources for manipulating the sound. They can be classified in two groups.
- Passive lighting (fixed and mobile) from spotlights which can be focused. The rays of these spotlights are unfiltered.
With this lighting, light produces stillness, and darkness produces sound.
- Active lighting from flames and projected images. The brightness produced by these sources is variable.
With this lighting, light produces sound, and darkness produces stillness.

Beschreibung des Environments

Electro Clips ist eine Installation für Ballett, die es Tänzern ermöglicht, mit den Elementen Licht und Ton unmittelbar zu interagieren.

Die visuellen und akustischen Zeichen, die das Ballett üblicherweise als vorgegeben benutzt, werden in „Electro Clips" von dem Tänzer Stephen Galloway, seiner Bewegung und Choreographie, selbst herbeigeführt und beeinflußt.

Es wird ein paralleles Environment aus Klang, Licht und Bewegung erzeugt. Der Tänzer übernimmt die Stelle des Dirigenten, indem er die wechselnden Funktionen der im Bühnenraum verteilten Sensoren zur Manipulation von Klang und Licht wie eine Klaviatur gebrauchen kann.

Die elektronische Steuerung

Im Bühnenboden sind Photoresistoren eingelassen, die in Abhängigkeit von der aktuellen Lichtmenge am jeweiligen Bühnenort einen variablen Widerstandswert ausgeben. Gemessen werden die daraus entstehenden, für das Publikum sichtbaren, veränderlichen Licht- und Schattenkompositionen, die im Bewegungsablauf des Tänzers entstehen sobald er in die Lichtkegel eintritt. Die jeweils verbleibende Lichtmenge wird in Form elektrischer Spannung an das Computersystem übertragen, digitalisiert und als MIDI-Information dem Audiosystem übermittelt.

Die Beleuchtung

„Electro Clips" verwendet unterschiedliche Arten von Lichtquellen zur Manipulation des Klangs. Sie lassen sich in zwei Hauptgruppen einteilen:
- die passive Beleuchtung (lokal und mobil) mittels fokussierbarer Scheinwerfer. Das sind Lichtquellen mit gleichmäßigem Strahlengang.
Licht erzeugt hier Stille und Dunkelheit erzeugt Klang.
- die aktive Beleuchtung mittels Flammen oder Bildprojektionen. Es sind Lichtquellen mit variabler Helligkeit.
Licht erzeugt hier Klang und Dunkelheit erzeugt Stille.
Bei der Ausleuchtung einer Szene mit passiven Lichtquellen sind es ausschließlich die Bewegungen des Tänzers, die die klanglichen Reaktionen des Environments hervorrufen. Ist die Szene aktiv beleuchtet, manipuliert der Tänzer eine bereits reagierende Klangkulisse. Die beim hellen Aufflammen eines Streichholzes abrupt und heftig reagie-

rende Klangkulisse wird im Verlauf des Abbrennens wieder ruhiger, mit dem Verlöschen wieder still. Die „Filter" (Bilder) der Projektoren – beim Diaprojektor austauschbar, jedoch statisch; beim Video-Beamer bewegt – modulieren mit ihrer Hell-Dunkel-Verteilung auf den Sensorfeldern differenzierte Klangkompositionen.

Das Audiosystem

Die 24 Kanäle eines Quadra 650 (3 x Sample Cell 2) sind mit verschiedenen Sounds (Stimmen) belegt und werden von einer MIDI-Steuereinheit verwaltet. Sie werden abhängig vom Ort auf der Bühne und von der gemessenen Lichtmenge aufgerufen und in variabler Lautstärke abgespielt.
Die dem Spannungsverlauf der Produktion angepaßte Komposition ist eine ständige Maximal-Orchestrierung. Die Soundtracks sind mit Einzelsounds, Textclustern, Rhythmen und Melodien belegt.
Das Klangerlebnis während der Performance wird durch den Umgang der Tänzer und der variablen Lichtquellen mit der Sensorik der Anlage generiert. Es handelt sich um eine interaktiv vorgenommene Selektion aus der laufenden Maximal-Orchestrierung.
Sechs unabhängige Lautsprechersysteme übertragen die selektierte Summe der Soundtracks dreidimensional in den Raum.

When lighting a scene with passive sources of light, the dancer's movements alone produce the environment's audio reactions. If the scene is lit actively, the dancer manipulates an acoustic background which is already reacting. The acoustic background, which reacts abruptly and violently to the bright flare of a match, becomes calmer as the match burns down and then quiet when it burns out. The projectors' "filters" (images) – exchangeable in the case of the slide projector, though static and mobile with the video projector – modulate differentiated audio compositions by means of their distribution of light and shade on the sensor fields.

Audio system

The 24-channels of a Quadra 650 (3 Sample Cell IIs) are assigned various sounds and are administered by a MIDI control unit. They are played back in relation to the specific location on the stage and the amount of light measured, and at different volumes.
The composition, adjusted to the voltage during the production, is a constant maximum orchestration. The sound tracks are assigned individual sounds, text clusters, rhythms and melodies.
The sound effects during the performance will be generated by the actions of the dancers and the variable light sources with regard to the system's sensors. This will be a selection made interactively from the running maximum orchestration.
Six independent loudspeaker systems will transmit the selected total sound tracks into the room in three-dimensions.

Musik: Peter Kuhlmann (D)
Licht: Louis Philippe Demers (CAN)
Programmierung: Louis Philippe Demers (CAN), Daniel Schmitt (D), Sven Thöne (D)
Produktion: Theater Am Turm, Frankfurt / Ars Electronica, Linz
Mit freundlicher Unterstützung des Forums für Informationstechnik GmbH, Paderborn

Music: Peter Kuhlmann (Germany)
Light: Louis Philippe Demers (Canada)
Programming: Louis Philippe Demers (Canada), Daniel Schmitt (Germany), Sven Thöne (Germany)
Production: Theater Am Turm, Frankfurt / Ars Electronica, Linz
With the welcome support of the Forum für Informationstechnik GmbH, Paderborn

Clip 1 (24:1)

- Ein einsames großes Spotlight bildet einen Lichtkegel auf der Bühne. "Stille"
- Ein Tänzer nähert sich dem Kegel
- Eine leichte Berührung des Lichtkegels öffnet durch die Verschattung auf dem Photoresister den Ersten von 24 Audiokanälen.
- Nach und nach greift der Tänzer stärker in den Lichtkegel hinein und vergrößert somit die Summe der zu hörenden Sounds. Die Soundabfolge ist linear, d.h. die Reihenfolge wie sich die Einzelsounds zu einer immer dicker werdenden Summe addieren und umgekehrt bleibt immer gleich.

Spotlight / Photoresister

Clip 2 (6:4)

1. Spotlights
1. Melodie — Flöte
2. Harmonie — ...
3. Bass — Rhythmus
4. Perkussion

- Drei weitere Spotlights im Hintergrund schalten zu.
- Die 24 möglichen Soundtracks werden auf die vier Sensoren verteilt.
- Dialog zwischen Flöte und Rhythmus mit heftigen Breaks.
- Die Breaks werden betont
- Auf dem Höhepunkt der Komposition stehen alle 4 Tänzer still

Clip 4 (2-12)

- Eine Videokamera filmt den Tänzer mit einem fokusierten Halogenstrahler. Close ups.
- Das Videobild wird von einem Beamer senkrecht von oben auf das Sensorenfeld projiziert.
- Die Hell-Dunkelkontraste des Videobildes triggern die Komposition sobald die senkrechten Spotlights ausgefadet sind.
- Der Tänzer betritt sein eigenes Videobild und überlappt mit seinen Verschattungen die hellen Bildstellen

Videobeamer / Kamera / Flächer

Clip 5 (1:24)

Flöte / Rhythmus

- Alle 24 Spotlights schalten an.
- Ein Tänzer im Sensorenfeld bestimmt den Rhythmus
- Andere Tänzer bestimmen die Fläche und die Melodie

FIESTA ELECTRA

Mega-Techno-Rave-into-Paradise-Party

FRIEDRICH GULDA AND HIS PARADISE BAND

ABSCHLUSSVERANSTALTUNG VON ARS ELECTRONICA
Sporthalle Linz
**Samstag, 25. Juni, 21.30 Uhr
open end**

Friedrich Gulda präsentiert „eine rockige, jazzige, soulige, lustbetont-also-wahrhaft-avantgardige, dancefloorige Elektro-Techno-Party der klassen Art"   FRIEDRICH GULDA

ARS ELECTRONICA, CONCLUDING EVENT
Sporthalle Linz
**Saturday, June 25, 9:30 PM
open end**

Friedrich Gulda presents „a rockin', jazzy, soulful, passionate-and-therefore-really-avant-garde, dance floor electro-techno party of the finest sort"   FRIEDRICH GULDA

Es treten auf / The musicians:
FRIEDRICH GULDA
AND HIS PARADISE BAND:
FRIEDRICH GULDA (piano, clavinova),
HARRY SOKAL (saxes),
MITCH WATKINS (guit.),
STEFAN MITTERBACHER (keyboard),
WAYNE DARLING (bass),
MICHAEL HONZAK (drums),
LAURINHO BANDEIRA (percussion),
KATHY SAMPSON, GINA CHARITO,
DORETTA CARTER (vocals)

PHILL EDWARDS (vocals)
KOFFI KOKO (dance)
FUMILAYO WEBER-JOHNSON (dance)

IBIZA AMNESIA DANCERS

DJ´s DOMINIQUE ANIMATOR +
JAIME SILVA

SDP Techno Paradise Sound and Light

INTELLIGENT / AMBIENT TECHNO

UNDERGROUND RESISTANCE, STATION ROSE, UNITED FREQUENCIES OF TRANCE live im Brucknerhaus, 24. Juni 1994.
live in the Brucknerhaus, June 24, 1994.

The theme of this year's Ars Electronica is, as is generally known, „Intelligent Ambiences". An apparent coincidence exists in that possibly the most interesting developments in the music of the techno movement, which resembles a healthy plant with the branches that have sprouted in the past five years, are termed „intelligent" and „ambient". These two branches, which are the newest, represent a notable turning point in the recent history of this electronic form of music. If techno was a music form used exclusively by DJs to hypnotize the masses from the turntable until relatively short time ago, ambient and intelligent have freed themselves of this duty and have developed into a musical art form.

„There's no techno tonight, we're playin' ambient." Well-known bouncer in front of the „XS" club in Frankfurt, every Sunday evening.

This style is consumed not only in clubs and at raves; large quantities are also taken at home in the form of CD/LP. Dancing generally bothers no one at ambient and intelligent parties, although the majority is there to experience the music at the required volume.

A similar phenomenon has already appeared in the history of popular culture. When Beat groups invented psychedelic music in the mid-60s, the nightclub managers were irritated at first because the guests just sat around and did not drink alcohol. Beer, normally held in the hand lovingly at concerts in the 80s, is being increasingly substituted by psychoactive soft drinks in the techno scene. And there is still another similarity: The prominent sound generators of psychedelic music, the sitar and the wah-wah guitar and their chirps and buzzes resemble the sounds produced by the mass-produced analog synths of the Japanese company Roland with their short filter attacks, long cut-off and resonance values with a modulated envelope amount. This trend away from compulsive hedonism, the turn to becoming involved in the creative process, to techno events with solos, to the realtime experience of a live performance

Das Thema der diesjährigen Ars Electronica ist, wie allgemein bekannt, „Intelligente Ambiente". Eine scheinbare Zufälligkeit besteht darin, daß die wohl interessantesten Entwicklungen der Musik der Technobewegung die mit ihren Verzweigungen in den letzten 5 Jahren einer gesunden Pflanze gleicht, „intelligent (-techno)" und „ambient (-techno)" genannt werden. Diese beiden jüngsten Äste markieren einen bemerkenswerten Wendepunkt in der jungen Geschichte dieser elektronischen Musik. War Techno (oder Tekkno, oder Tekno) bis vor nicht allzu langer Zeit fast ausschließlich Musik, die DJs dazu diente, die Massen vom Plattenteller aus zu hypnotisieren, so haben sich ambient und intelligent von dieser Pflicht losgelöst, und zu einer musikalischen Kunstform entwickelt.

„Heute gibt's kein' Techno, heute is' ambient." Bekannter Türsteher vor dem Frankfurter Club „XS", jeden Sonntag abend.

Dieser Stil wird nicht nur in Clubs und auf Raves, sondern in großem Maß zu Hause in CD/LP-Form zu sich genommen; bei ambient- und intelligent-Parties stört es zwar niemanden, wenn getanzt wird, die meisten sind jedoch da, um die Musik in der nötigen Lautstärke zu erleben.

Ein ähnliches Phänomen gab es schon einmal in der Geschichte der Populärkultur. Als Mitte der 60er Jahre Beatgruppen die psychedelische Musik erfanden, waren Nachtklubmanager zunächst irritiert, daß die Gäste bloß herumsaßen und keinen Alkohol konsumierten. Bier, vom Konzertgänger der 80er noch liebevoll meist stehend in der Hand gehalten, wird in der Technoszene immer mehr durch psychoaktive Softdrinks substituiert. Und noch eine Ähnlichkeit: die markanten Soundgeneratoren der Psychedelischen Musik, die Sitar und die Wah-Wah-Gitarre, ähneln mit ihren zirpenden und surrenden Klängen den Sounds der bei ambient und intelligent gerne verwendeten Fließband-Analogsynthies der japanischen Firma Roland mit ihren kurzen Filterattacks, langen Cut-Off- und Resonanzwerten bei moduliertem Envelope-Amount.

Diesen Trend Weg-vom-Zwangshedonismus, die Hinwendung zum Hineindenken in den Schaffensprozess, zum konzertanten Techno-Event, zum Echtzeiterlebnis einer Live-Performance, beschreibt MAD MIKE, der Kopf der intelligent-Pioniere UNDERGROUND RESISTANCE aus Detroit, in einem Fax an Station Rose vom 28.3.1994:
UNDERGROUND RESISTANCE'S LIVE SHOW IS BASED ON U.R.'S TOTAL COMMITMENT TO EXPERIMENTAL MUSIC. OUR

CONCEPT OF MAN-MACHINES AND TECHNOLOGY IS TESTED EVERY TIME WE PLAY LIVE, IN OUR EXPERIMENTS ON STAGE WE PROBE OUR THEORY THAT SOUND IS AN UNRECOGNIZED LIFEFORM THAT MAN HAS LITTLE TO NO KNOWLEDGE OF. THE MEN OF U.R. TEAMED WITH THEIR UNDEPENDABLE ANCIENT OLD ANALOG EQUIPMENT ALONG WITH THE ADDED UNPREDICTABILITY OF A LIVE SHOW AND AUDIENCE, LEADS TO A ONE-TIME ONLY EXPERIENCE FOR US AS MUSICIANS AND FOR THE AUDIENCE TOO, WHICH IN THE END WE HOPE WILL BROADEN ALL OF OUR UNDERSTANDINGS OF SONIC ACTIVITY AND HOW IT AFFECTS US.   –
OUT –   MAD MIKE

Im sinnlichen Umgang mit Technologie werden Analogsynthies aus den frühen 80ern in MIDI-Schleifen mittels moderner Samplingtechnologie eingebunden. Der Techno-Live-Künstler übersiedelt für die Performance mit seinem Studio auf die „Bühne" und erstellt dort in Echtzeit neue Varianten seiner Sequenzen, neue Klangfarben, zeitliche Strukturen und Loops.

A NUMBER OF THEMES ARE PRE-DETERMINED BEFORE THE PERFORMANCE AND ELEMENTS ASSIGNED TO EACH MUSICIAN. IN ADDITION EACH MUSICIAN HAS HIS OWN PRE-PROGRAMMED SEQUENCES WHICH ARE NOT REVEALED TO THE OTHERS BEFORE THE PERFORMANCE. CERTAIN MACHINES ARE ALSO SET UP TO BEHAVE IN A TOTALLY RANDOM FASHION. UNITED FREQUENCIES OF TRANCE / Dominic Woosey

Station Rose erweitert den Live-Begriff um die Multimediakonstante und ergänzt die Performance mit Telepresence-Feeling durch globale Echtzeitvernetzung: Zur Maßeinheit von B.P.M (beats per minute) kommt eine neue Maßeinheit für Echtzeitkunst dazu R.P.M (realities per minute).

BEI IHREN LIVE-PERFORMANCES ERZEUGT STATION ROSE MITTELS SOUND- UND PROJEKTIONSFLÄCHEN EINEN „VIRTUELLEN RAUM IM RAUM". DIESER BESTEHT AUSSCHLIESSLICH AUS LICHT (BEAMS) UND PROJEKTIONSFLÄCHEN ,

was described by MAD MIKE, the head of the intelligent pioneers UNDERGROUND RESISTANCE from Detroit, in his fax of 3/28/1994 to Station Rose:
UNDERGROUND RESISTANCE´S LIVE SHOW IS BASED ON U.R.´S TOTAL COMMITMENT TO EXPERIMENTAL MUSIC. OUR CONCEPT OF MAN-MACHINES AND TECHNOLOGY IS TESTED EVERY TIME WE PLAY LIVE, IN OUR EXPERIMENTS ON STAGE WE PROBE OUR THEORY THAT SOUND IS AN UNRECOGNIZED LIFEFORM THAT MAN HAS LITTLE TO NO KNOWLEDGE OF. THE MEN OF U.R. TEAMED WITH THEIR UNDEPENDABLE ANCIENT OLD ANALOG EQUIPMENT ALONG WITH THE ADDED UNPREDICTABILITY OF A LIVE SHOW AND AUDIENCE, LEADS TO A ONE-TIME ONLY EXPERIENCE FOR US AS MUSICIANS AND FOR THE AUDIENCE TOO, WHICH IN THE END WE HOPE WILL BROADEN ALL OF OUR UNDERSTANDINGS OF SONIC ACTIVITY AND HOW IT AFFECTS US.   – OUT –   MAD MIKE

In this method of working with technology through the senses, analog synths from the early 80s are connected in MIDI loops by means of modern sampling technology. The live techno artist moves his studio to the „stage" for the performance and creates new variations on his sequences, new tones, temporal structures and loops in realtime.

A NUMBER OF THEMES ARE PRE-DETERMINED BEFORE THE PERFORMANCE AND ELEMENTS ASSIGNED TO EACH MUSICIAN. IN ADDITION EACH MUSICIAN HAS HIS OWN PRE-PROGRAMMED SEQUENCES WHICH ARE NOT REVEALED TO THE OTHERS BEFORE THE PERFORMANCE. CERTAIN MACHINES ARE ALSO SET UP TO BEHAVE IN A TOTALLY RANDOM FASHION. UNITED FREQUENCIES OF TRANCE / Dominic Woosey

Station Rose expands the „live" concept by its multimedia constants and complements the performance with telepresence feeling via realtime global networking:
DURING THEIR LIVE PERFORMANCES, STATION ROSE CREATES „A VIRTUAL ROOM WITHIN A ROOM" BY MEANS OF SOUND AND PROJECTION SURFACES. THIS ROOM CONSISTS EXCLUSIVELY OF BEAMS AND PROJECTION SURFACES ONTO WHICH VISUALS ARE THROWN. THESE IMAGES ARE DOWNLOADED AND EDITED LIVE ON THE COMPUTER. THE LIVE MUSIC CONSISTS OF SOUND SAMPLES FROM THE DIGITAL ARCHIVE OF STATION ROSE WHICH ARE COMBINED INTO COMPOSITIONS

ON THE SEQUENCER. WITH A MODEM, A REAL-TIME CONNECTION IS CREATED WITH THE E-MAIL NETWORK „THE WELL" IN CALIFORNIA, OF WHICH STATION ROSE HAS BEEN A MEMBER SINCE 1991. THIS NETWORK ALLOWS USERS AROUND THE GLOBE TO PARTICIPATE IN THE EVENT. CHANGES WILL TAKE PLACE DURING THE PERFORMANCE, THEREBY CHANGING THE AESTHETICS OF THE VIRTUAL LIGHT SPACE; INFORMATION WILL FLOW THROUGH THE MAILBOX ONTO THE WALLS ILLUMINATED WITH COMPUTER EMISSIONS. THE NETWORKING PROCESS CAN BE OBSERVED ON THE PROJECTION SURFACES AS A FURTHER OPTICAL ELEMENT.

STATION ROSE / Gary Danner & Elisa Rose
gunafa@well.sf.ca.us

WORAUF VISUALS GEWORFEN WERDEN, DIE LIVE AM COMPUTER DOWNGELOADET UND EDITIERT WERDEN. DIE LIVEMUSIK BESTEHT AUS SOUNDSAMPLES AUS DEM DIGITALEN ARCHIV DER STATION ROSE, DIE AM SEQUENZER ZU KOMPOSITIONEN VERBUNDEN WERDEN. MITTELS MODEM WIRD LIVE EINE ECHTZEIT-VERNETZUNG MIT DEM E-MAIL NETZ „THE WELL" IN KALIFORNIEN HERGESTELLT, DESSEN MITGLIED STATION ROSE SEIT 1991 IST UND DAS USER GLOBAL AN DEM

EVENT TEILNEHMEN LÄSST. VERÄNDERUNGEN TRETEN WÄHREND DER DAUER DER PERFORMANCE AUF UND VERÄNDERN DIE ÄSTHETIK DES VIRTUELLEN LICHTRAUMES; INFORMATION FLIESST ÜBER DIE MAILBOX AUF DIE VON COMPUTEREMISSIONEN BESTRAHLTEN WÄNDE. DIE VERNETZUNG IST ALS WEITERES OPTISCHES ELEMENT AUF DEN PROJEKTIONSFLÄCHEN MITZUVERFOLGEN...

STATION ROSE / Gary Danner & Elisa Rose
gunafa@well.sf.ca.us.

DIE ARS ELECTRONICA
15 Jahre Festival für Kunst, Technologie und Gesellschaft
15 Years of the Festival of Art, Technology and Society

AUSSTELLUNG IN DER OÖ. LANDESGALERIE, MUSEUM FRANCISCO CAROLINUM
EXHIBITION IN THE UPPER AUSTRIAN LANDESGALERIE, MUSEUM FRANCISCO CAROLINUM

„Was ich jedoch aus den Veröffentlichungen über vorangegangene Treffen gelesen habe und während des letzten Treffens (von Ars Electronica) selbst erleben durfte, hat mich in der Überzeugung bestärkt, daß wir hier mit einem Zentralthema der Gegenwart konfrontiert sind."

Vilém Flusser

"However, what I could gather from publications about previous meetings, together with what I was able to experience for myself at Ars Electronica, has strengthened my conviction that what we are confronted with here is a central issue of the present time."

Vilém Flusser.

Die – nunmehr bereits 15jährige – Geschichte von Ars Electronica, des konsequent auf die Zukunft ausgerichteten Linzer Festivals für Kunst, Technologie und Gesellschaft, das sich dem kreativen Umgang mit den neuen elektronischen Medien widmet, ist – gleichsam aus der Eigendefinition heraus – eine sich permanent in ihren Grundstrukturen verändernde.

Ars Electronica legt sich weder hinsichtlich des Zielpublikums noch der fachlichen Zuordnung Einschränkungen auf; das kreative, auf die Weiterentwicklung in der Zukunft ausgerichtete Element blieb und bleibt stets das hauptsächliche Anliegen: So veranstaltet Ars Electronica zum einen hochkarätige Fachsymposien mit internationalen Experten, zum anderen Klangwolken, die bis zu 150.000 Besucher im Linzer Donauraum konzentrierten. Ars Electronica präsentierte Computerkunst als Bildvorstellungen, als Klangräume, als komplexe Rauminstallationen, aber auch als künstlerische Eingriffe in den öffentlichen Raum. Immer wieder wurden neue Formen der Inszenierungen von Farben, Formen und Klängen gezeigt. So überschritten die Veranstaltungen des Linzer Festivals beständig Grenzen der Zuordnungen und der Erwartungen: Die Kunst drang als Wirkungskraft immer prägnanter in neue gesellschaftliche Kontexte vor.

Die Ausstellung in der O.Ö. Landesgalerie versucht daher nicht die (unlösbare) Aufgabe, eine Art „Hitliste" der interessantesten, nachwirkendsten, aufregendsten, diskutiertesten, umfangreichsten, komplexesten, prominentesten u.s.w. Veranstaltungen aus der Geschichte des vielgesichtigen Festivals zusammenzustellen, sondern möchte durch konzentrierte Schwerpunktsetzungen jene Bereiche herausstreichen, in denen Ars Electronica prägend wirksam wurde. Die Auswahl hat natürlich Zitatcharakter und bezieht sich vor allem auf die Kraft des jeweiligen Erlebnisses. Durchwegs alle Produktionen von Ars Electronica waren ja auf die Einmaligkeit eines genau orts- und zeitspezifischen Erlebnisses hin ausgerichtet und wurden zudem in den umfangreichen Katalogpublikationen sehr detailliert dokumentiert. In diesem Sinne bemüht sich die

Ars '82: Charlotte Moorman und Otto Piene: "Sky Kiss"

Ars Electronica is a festival for art, technology and society. For fifteen years now it has maintained a consistently forward-looking approach and dedicated itself to the creative use of the new electronic media. In accordance with this self-definition, so to speak, the history of the festival is one of continual change from the foundations up. Ars Electronica does not burden itself with any limitations regarding target audience or technical distinctions; the creative element, dedicated to further development in the future, has always remained the primary concern: so on the one hand Ars Electronica hosts a high-powered, specialist symposium with international experts; on the other, it attracts up to 150.000 visitors to the

Danube Park in Linz for the Klangwolke (Sound Cloud). Ars Electronica has presented computer art in the form of picture exhibitions, sound-spaces, complex space installations, and even as artistic interventions in public spaces. Time and again the festival has presented new ways of arranging colours, shapes and sounds, allowing festival events to consistently transcend the limits of categorisation as well as the limits of expectation: Art has penetrated ever further, ever more appropriately, into new social contexts, as a force to be reckoned with.

The exhibition at the Upper Austrian Landesgalerie is not therefore intended to perform the (impossible) task of putting together a kind of "hit list" of the most interesting, most influential, most exciting, most discussed, most comprehensive, most complex, most noticeable (and so forth) events from the history of the multi-faceted festival, but to use concentrated emphasis to underline those areas in which Ars Electronica has come to exert a formative influence. Naturally the selection gives the impression of a quotation, and above all, makes reference to the power of the experience in question. All Ars Electronica productions have after all been tailored to the uniqueness of an experience specific to one particular place and time, and were, moreover, documented in great detail in the comprehensive catalogue books. In this sense and with the help of electronic media, the exhibition strives to convey the fifteen-year history of Ars Electronica not as the recreation of an historical event from the distant past, but as an intellectual and sensory experience.

This exhibition is divided into the following areas:

1. Information:

- Basic information about previous Ars Electronica festivals, their overall themes, the people responsible for the programme of events, the projects that were presented, the symposia, concerts and performances, are all featured in a moving body of text with individual photographs.
- A specially designed overview of the information is also available on a CD-ROM, "15 Years of Ars Electronica", put together by Station Rose, which gives the visitor interactive access to make his or her own way through information stored in the form of text, still pictures and video (see Vol. 1, p. 201).
- Every year, the highlights of the Ars Electronica festival are summarised in a TV documentary. This material, which can already be deemed historic, gives a very direct and vivid insight not

Ausstellung mit Hilfe elektronischer Medien darum, die 15jährige Geschichte von Ars Electronica nicht als distanziert historisches Ereignis nachzuvollziehen, sondern als geistiges und sinnliches Erlebnis zu vermitteln.

Die Ausstellung gliedert sich in folgende Bereiche:

1. Information:

– Grundsätzliche Informationen über die bisherigen Ars Electronica-Festivals, die Hauptthemen, die Programmgestalter, die präsentierten Projekte, die Symposien, Konzerte und Inszenierungen werden über ein durchlaufendes Schriftband, in das einzelne Fotografien integriert sind, vermittelt.

– Einen speziell gestalteten Informationsüberblick bietet zudem eine von der Station Rose erarbeitete CD-Rom zum Thema „15 Jahre Ars Electronica", die dem Besucher

Goldene Nica Prix Ars Electronica '87: "Luxo jr." von John Lasseter

– über interaktive Zugriffsmöglichkeiten auf die in Schrift-, Bild- und Filmform gespeicherten Daten – ermöglicht, sich einen eigenen Informationsweg zu schaffen (vgl. hierzu Band 1, S. 201)

– Eine filmische Zusammenfassung der wichtigsten Programmpunkte der einzelnen Ars Electronica-Festivals wurde in den speziell zu jedem Festival gestalteten Fernsehdokumentationen erarbeitet. Dieses schon als historisch zu bezeichnende Filmmaterial bietet somit einen sehr direkten und lebendigen Einblick nicht nur in die Programmabfolge, sondern auch in die Stimmungen der Festivals. Im Rahmen der Ausstellung werden diese Dokumentationen in einer permanenten Abfolge zu sehen sein; es besteht aber auch die Möglichkeit, sich einzelne Dokumentationsfilme auf Wunsch separat zu betrachten.

– Im Festsaal des O.Ö. Landesmuseums wird durch eine Videogroßprojektion eine Kinosituation geschaffen: Hier werden sowohl alle Preisträgerarbeiten des Prix Ars Electronica seit Bestehen dieses Preises aus der Sparte der Computeranimation wie auch speziell für

Ars Electronica geschaffene Videobänder präsentiert.

2. Relikte

Aus der großen Fülle an künstlerischen Projekten, die im Rahmen von Ars Electronica verwirklicht werden konnten, wurden für diese Ausstellung einige markante ausgewählt: Sie stehen als Beispiele für die große Vielfalt sowie für die beständigen Grenzüberschreitungen. Als Hinweise auf diese Projekte werden einige Relikte, also Materialreste – kombiniert mit kurzen filmischen Sequenzen bzw. Fotografien der Projektsituation – gezeigt.

3. Tunnel der Feststellungen und Visionen

Im Verlauf der einzelnen Ars-Electronica-Festivals wurden vielfach, zumeist bei den wissenschaftlichen Fachsymposien, sehr prägnante Äußerungen über die gesellschaftliche, technologische und künstlerische Problematik der elektronischen Medien formuliert. Neben aktuellen Erläuterungen beziehen sich diese Feststellungen auch immer wieder auf zukünftige Visionen oder Erwartungen.

Ars '79: SPA 12, einer der ersten elektronischen Roboter kam aus den USA nach Linz

Bei der Ausstellung werden solche markante Formulierungen aus der 15jährigen Ars-Electronica-Geschichte thematisch geordnet und somit in neue historische Zusammenhänge gebracht. Zukunftsprognosen, die teilweise inzwischen in überraschend präziser Weise Realität geworden sind, lassen sich so überprüfen und verraten zudem viel über den geistigen Umgang mit der Computerwelt in den vergangenen 15 Jahren.

4. Klangwolke

Auch wenn die Linzer Klangwolke inzwischen nicht mehr Teil von Ars Electronica ist, sondern nun alljährlich im Rahmen des Brucknerfestes realisiert wird, ist sie dennoch aufs engste mit der Geschichte von Ars Electronica verbunden. Bereits die erste Ars Electronica im Jahr 1979 präsentierte als gesellschaftlichen Höhepunkt des Festivals das spezielle „Klangwolkenerlebnis" eines raumgreifenden Konzertes mit entsprechender optischer Umsetzung in der freien Natur. Dieses besondere Klangwolkenerlebnis wird bei der Ausstellung über eine interaktive Klanginstallation nachvollziehbar gemacht.

only into the programme of events itself but also into the atmosphere of the festival. These documentaries will be shown in continuous rotation as part of the exhibition; however it will also be possible to view them individually on request.
- A cinema setting will be installed in the ballroom of the Upper Austrian Landesmuseum using video projection facilities. All the prize-winning computer animations from the eight year history of the Prix Ars Electronica will be featured along with their creators, as will videotapes created especially for Ars Electronica.

2. Relics

A few of the great many artistic projects realised at Ars Electronica have been selected for the exhibition: they testify to the considerable diversity and the breaking of boundaries the festival entails. The exhibition will pay tribute to these projects in the form of relics – material remains – combined with short video sequences and photographs.

3. Tunnel of observations and visions

In the course of the many Ars Electronica festivals, particularly at the scientific symposia, some very perceptive things have been said about social, technological and artistic issues relating to the electronic media. Besides topical remarks about the contemporary state of affairs, these observations often relate to visions, or expectations, of the future.
Memorable statements of this kind, excerpted from the whole 15-year history of Ars Electronica, have been arranged thematically for this exhibition and thereby given new historical contexts and associations. This can be used to test earlier predictions for the future, which in some cases have proved surprisingly accurate, and furthermore, much is revealed about the intellectual use of the computer over the past 15 years.

4. Klangwolke (Sound Cloud)

Even though the Linz Klangwolke is no longer a part of Ars Electronica but the opening event of the annual Brucknerfest, it is still closely associated with the history of Ars Electronica. From the very first Ars Electronica in 1979, the festival's social highlight was the "Sound Cloud experience", consisting of a large-scale concert with the appropriate optical interpretation, out in the open air. The exhibition will recreate this special "Klangwolke" experience by means of an interactive sound installation.

5. Interactive television

One of the particularly important Ars Electronica innovations is its involvement with interactive television, an experiment which started in the early days of the festival and has continued ever since. In addition to the current development of such a television project in the Linz Brucknerhaus concert hall (see p. 12), the exhibition in the Upper Austrian Landesgalerie is presenting a room installation by the Linz Stadtwerkstatt ("City Workshop"; cultural centre) featuring projects which have been carried out in the past by the Stadtwerkstatt or by Van Gogh TV.

6. The Prix Ars Electronica 1994

The exhibition of the 15 years of Ars Electronica is not limiting itself to displaying history; as a reference to the latest artistic forms of electronic media involvement (complementing the other sites of the Ars Electronica exhibition), it will present a selection of works that have been awarded prizes in the Computer Graphics and Interactive Art categories of this year's Prix Ars Electronica.

PETER ASSMANN

Responsible for concept and coordination:
PETER ASSMANN

Exhibition organisation: Upper Austrian Landesgalerie, ORF Upper Austrian Branch and the Brucknerhaus Linz.
Exhibition design: Sepp Auer and Robin Hood Inc.

We would like to express our thanks to the following companies, without whose generous support this exhibition project could not have been realised: SHARP / PHILIPS / EBG (Elektro Bau AG) COMPUTER CORNER / SILICON GRAPHICS

5. Interaktives Fernsehen

Als eine der speziell wichtigen Innovationen von Ars Electronica ist die sehr früh begonnene und permanent fortgesetzte Auseinandersetzung mit interaktiven Fernsehprojekten anzusehen. Neben der aktuellen Realisation eines solchen Fernsehprojektes im Linzer Brucknerhaus (vgl. S 12) präsentiert die Ausstellung in der O.Ö. Landesgalerie eine Rauminstallation der Stadtwerkstatt Linz, in der auf die vergangenen Projekte – sei es von dieser Institution oder von Van-Gogh-TV – hingewiesen wird.

6. Prix Ars Electronica 1994

Die 15-Jahre-Ars-Electronica-Ausstellung in der Landesgalerie beschränkt sich jedoch nicht allein darauf, Geschichte zu zeigen: Im Sinne eines Hinweises auf die aktuellsten künstlerischen Formen der Auseinandersetzung mit elektronischen Medien werden – in Ergänzung zu den anderen Ausstellungsorten von Ars Electronica – eine Auswahl der im heurigen Jahr beim Prix Ars Electronica ausgezeichneten Projekte aus den Sparten „Computergraphik" und „Interaktive Projekte" präsentiert.

PETER ASSMANN

Konzeptions- und Koordinationsverantwortung:
PETER ASSMANN

Ausstellungsrealisation: O.Ö. Landesgalerie, ORF-Landesstudio Oberösterreich, Brucknerhaus Linz.
Ausstellungsgestaltung: Sepp Auer und Robin Hood Inc.

Für die großzügige Unterstützung, ohne die eine Realisation dieses Ausstellungsprojektes nicht möglich gewesen wäre, danken wir folgenden Firmen: SHARP / PHILIPS / EBG (Elektro Bau AG) COMPUTER CORNER / SILICON GRAPHICS

Ars '88: Philosophische Tafelrunde (v.l.n.r.): Peter Gente, Hannes Böhringer, Vilém Flusser, Peter Weibel, Jean Baudrillard, Friedrich Kittler, Heinz von Foerster

PRIX ARS ELECTRONICA 94

Wettbewerb für Computergraphik, -animation, -musik und Interaktive Kunst
Contest for interactive art and computer graphics, animation and music

In den 15 Jahren des Bestehens von Ars Electronica hat sich das Linzer Festival für Kunst, Technologie und Gesellschaft im Siebenjahreszyklus erneuert und verbreitert. 1987 kommt zum Festival der vom Österreichischen Rundfunk (ORF) initiierte Wettbewerb: der Prix Ars Electronica, und 1993/94 das Ars-Electronica-Center, das die Stadt Linz zur Zeit am Donauufer errichtet, und das ab 1996 als Museum der Zukunft und Ort des Lernens und Experimentierens mit neuen Medien offen stehen wird.

Während Ars Electronica mit jährlich wechselnden Themen künftige Entwicklungen im kulturellen Kontext zur Diskussion stellt, ist der Prix Ars Electronica das kontinuierliche Präsentationsforum von Produktionen im Spannungsfeld von Kunst und Technologie. Damit hat Ars Electronica seit 1987 ihre Basis auf zwei Standbeine gestellt.

War der Prix Ars Electronica in der Anfangsphase als internationaler Wettbewerb für Computerkünste konzipiert, ist er mittlerweile offen für Kunst, Wissenschaft, Forschung und für Produktionen aus der Unterhaltungsbranche. Ziel des Prix Ars Electronica ist es, hervorragende Leistungen auszuzeichnen und damit ein Signal in Richtung Zukunft zu setzen.

Ermöglicht wird der Prix Ars Electronica 94 durch das Sponsoring der Kapsch AG und durch die Förderungen der Stadt Linz und des Landes Oberösterreich. Preisgelder mit einer Gesamtdotation von 1,25 Mio. Schilling und die Verbindung mit dem international angesehenen Festival machen den Prix Ars Electronica zum jährlichen Pflichttermin für jene, die mit Hilfe des Computers neue Gestaltungsmöglichkeiten erforschen.

In the 15 years of the Ars Electronica, this festival in Linz for art, technology and society has been renewed and expanded in seven-year cycles. In 1987, the contest initiated by the Österreichische Rundfunk (the Austrian broadcasting corporation, ORF), the Prix Ars Electronica, was made a part of the festival, as will happen with the Ars Electronica Center in 1993/1994. This building, now being constructed by the city of Linz on the bank of the Danube, will open as a museum of the future and a place to learn and experiment with new media in 1996.

While the Ars Electronica's changing annual themes initiate discussion of future developments in a cultural context, the Prix Ars Electronica is the continuous forum for the presentation of productions in the area of tension between art and technology. Therefore, the Ars Electronica has had a dual foundation since 1987.

Although the Prix Ars Electronica was conceived as an international contest for computer art at the beginning, it has been opened to art, science, research and productions from the entertainment sector. The goal of the Prix Ars Electronica is to acknowledge outstanding achievements, thereby giving new impulses towards the future.

The Prix Ars Electronica 94 is made possible by funding from the Kapsch AG, the city of Linz and the province of Upper Austria. Prize money amounting to ATS 1.25 million and the tie to the internationally renowned festival makes the Prix Ars Electronica an annual must for those who are exploring the new possibilities of computer design.

Christa Sommerer (Österreich) /
Laurent Mignonneau (Frankreich):
A-VOLVE
Goldene Nica für Interaktive Kunst

Dennis Muren/Mark Dippé/ILM (USA):
JURASSIC PARK
Goldene Nica für Computeranimation

PRIX ARS ELECTRONICA GALA

VERLEIHUNG DER GOLDENEN NICA

Mittwoch, 22. Juni, 20.30 – 22.00 Uhr
Live-Übertragung in ORF 2 und 3SAT; ORF, Studio 3
Einladung erforderlich

Die Prix Ars Electronica Gala ist die Preisverleihung im Prix Ars Electronica 94.
Fünf Goldene Nicas und acht Auszeichnungen werden in diesem Jahr vergeben.
Die Goldenen Nicas erhalten Dennis Muren/Mark Dippé/ILM, USA, Marc Caro/Midi Minuit, Frankreich, (Computeranimation), Michael Joaquin Grey, USA (Computergraphik), Christa Sommerer, Österreich und Lauren Mignonneau, Frankreich (Interaktive Kunst) und Ludger Brümmer, Deutschland (Computermusik).
Die Auszeichnungen erhalten Eric Coignoux /Mikros Image, Frankreich, Maurice Benayoun/Z.A Productions, Frankreich (Computeranimation), Keith Cottingham, USA, John Kahrs/Sky Productions, USA (Computergraphik), TRANSIT, Österreich, Loren/Rachel Carpenter/Cinematrix, USA (Interaktive Kunst) und Ake Parmerud, Schweden und Jonathan Impett, Großbritannien (Computermusik).
Die Prix-Ars-Electronica-Gala 1994 wird von Mercedes Echerer präsentiert und von 21.00 – 21.55 Uhr live in ORF2 und im Satellitenprogramm 3SAT europaweit übertragen.

PRESENTATION OF THE GOLDEN NICA

Wednesday, June 22, 8:30 – 10:00 PM
Live broadcast on ORF 2 and 3SAT; ORF, Studio 3
Invitation required

The presentation of prizes from the Ars Electronica 94 will take place at the Prix Ars Electronica Gala.
Five Golden Nicas and eight other awards will be presented this year.
The Golden Nicas will be awarded to Dennis Muren/Mark Dippé/ILM, USA, Marc Caro/Midi Minuit, France, (computer animation), Michael Joaquin Grey, USA, (computer graphics), Christa Sommerer, Austria and Laurent Mignonneau, France (interactive art) and Ludger Brümmer, Germany (computer music).
Awards will be presented to Eric Coignoux/Mikros Image, France, Maurice Benayoun/Z.A. Productions, France (computer animation), Keith Cottingham, USA, John Kahrs/Sky Productions, USA (computer graphics), TRANSIT, Austria, Loren/Rachel Carpenter/Cinematrix, USA (interactive art) and Ake Parmerud, Sweden and Jonathan Impett, Great Britain (computer music).
The Prix Ars Electronica Gala 1994 will be hosted by Mercedes Echerer and broadcast throughout Europe from 9:00-9:55 P.M. on ORF2 and the satellite program 3SAT.

PRIX ARS ELECTRONICA KÜNSTLERFORUM

10.00 – 18.00 Uhr
ORF, Studio 3
Eintritt frei

Das Prix-Ars-Electronica-Forum stellt die Preisträger mit ihren ausgezeichneten Arbeiten in Vorträgen und Diskussionen vor.

10.00 AM – 6.00 PM
ORF, Studio 3
No admission charge

The Prix Ars Electronica Forum will present the prize winners and their works in lectures and discussions.

Donnerstag, 23. Juni 1994

COMPUTERANIMATION/COMPUTERGRAPHIK

10:00 – 10:45	MAURICE BENAYOUN Z.A Productions, Frankreich: Ils sont tous là! (Les QUARXS)
10:45 – 11:30	ERIC COIGNOUX, MIKROS IMAGE, Frankreich: No Sex!
11:30 – 12:15	GEORGES BERMANN, MIDI MINUIT, Frankreich: K.O. Kid
12:15 – 1:00	MARK DIPPÉ, IL&M, USA: Jurassic Park
1:00 – 2:00	Midday break
2:00 – 2:45	JOHN KAHRS, SKY PRODUCTIONS, USA: Supercluster
2:45 – 3:30	KEITH COTTINGHAM, USA: "Fictitious Portraits/Self Portrait"
3:30 – 4:15	MICHAEL JOAQUIN GREY, USA: Jelly Life
4:15 – 4:45	Coffee break
4:45 – 6:00	Discussion with the day's speakers on the topic of "Possibilities, Prospects and Limitations of Computer Graphics in Art, TV, Advertising and Film"

Freitag, 24. Juni 1994

COMPUTERMUSIK/INTERAKTIVE KUNST

10.00 – 10.45	AKE PARMERUD, Sweden: Strings&Shadows
10.45 – 11.30	JONATHAN IMPETT, Great Britain: Mirror-Rite
11.30 – 12.15	LUDGER BRÜMMER, FOLKWANG COLLEGE, Germany: The Gates of H.
12:15 – 1:00	Discussion of the topic "New Romanticism in Computer Music?"
1:00 – 2:00	Midday break
2:00 – 2:45	HEIDI GRUNDMANN/TRANSIT e.V., Austria: Realtime
2:45 – 3:30	LOREN AND RACHEL CARPENTER, CINEMATRIX™, USA: Audience Participation
3:30 – 4:15	CHRISTA SOMMERER, Austria and LAURENT MIGNONNEAU, France: A-Volve
4:15 – 4:45	Coffee break
4:45 – 6:00	Discussion with the speakers on the topic "Interactivity on the Search for Contents?"

PRIX ARS ELECTRONICA AUDIO AND VIDEO SHOP

Thursday, June 23 to Saturday, June 25, 1994, Daily, 10:00 AM – 6:00 PM
No admission charge

In addition to a selection of the best works of the Prix Ars Electronica 94, the visitor can put together his or her own program from the 1,584 submissions.

Donnerstag, 23. Juni 1994

COMPUTERANIMATION/COMPUTERGRAPHIK

10.00 – 10.45	MAURICE BENAYOUN, Z.A PRODUCTIONS, Frankreich: Ils sont tous là! (Les QUARXS)
10.45 – 11.30	ERIC COIGNOUX, MIKROS IMAGE, Frankreich: No Sex!
11.30 – 12.15	GEORGES BERMANN, MIDI MINUIT, Frankreich: K.O. Kid
12.15 – 13.00	MARK DIPPÉ, IL&M, USA: Jurassic Park
13.00 – 14.00	Mittagspause
14.00 – 14.45	JOHN KAHRS, SKY PRODUCTIONS, USA: Supercluster
14.45 – 15.30	KEITH COTTINGHAM, USA: „Fictitious Portraits/Self Portrait"
15.30 – 16.15	MICHAEL JOAQUIN GREY, USA: Jelly Life
16.15 – 16.45	Kaffeepause
16.45 – 18.00	Diskussion mit den Tagesvortragenden zum Thema „Möglichkeiten, Perspektiven und Grenzen der Computergraphik in Kunst, TV, Werbung und Film

Freitag, 24. Juni 1994

COMPUTERMUSIK/INTERAKTIVE KUNST

10.00 – 10.45	AKE PARMERUD, Schweden: Strings & Shadows
10.45 – 11.30	JONATHAN IMPETT, Großbritannien: Mirror-Rite
11.30 – 12.15	LUDGER BRÜMMER, FOLKWANG HOCHSCHULE, Deutschland: The Gates of H.
12.15 – 13.00	Diskussion zum Thema „Neue Romantik in der Computermusik?"
13.00 – 14.00	Mittagspause
14.00 – 14.45	HEIDI GRUNDMANN/TRANSIT, e.V., Österreich: Realtime
14.45 – 15.30	LOREN UND RACHEL CARPENTER, CINEMATRIX™, USA: Audience Participation
15.30 – 16.15	CHRISTA SOMMERER, Österreich und LAURENT MIGNONNEAU, Frankreich: A-Volve
16.15 – 16.45	Kaffeepause
16.45 – 18.00	Diskussion mit den Tagesreferenten zum Thema „Interaktivität auf der Suche nach Inhalten?"

PRIX ARS ELECTRONICA AUDIO- UND VIDEOTHEK

Donnerstag, 23. bis Samstag, 25. Juni 1994 täglich 10.00 – 18.00 Uhr
Eintritt frei

Neben einer Auswahl der besten Arbeiten im Prix Ars Electronica 94 kann sich der Besucher aus sämtlichen 1.584 Einreichungen sein eigenes Wunschprogramm zusammenstellen.

COMPUTER und SPIELE

computer and games

Interaktive Ausstellung prämierter Schülerarbeiten
Interactive Exhibit of Premiered Student Works

Ergebnisse des bundesweiten Wettbewerbes des Österreichischen Kultur-Service in Zusammenarbeit mit dem Bundesministerium für Unterricht und Kunst und Ars Electronica.

Der Umgang mit Computertechnologien ist speziell für Jugendliche eine Selbstverständlichkeit. Es gibt bereits eine eigene „Demo-Group-Szene" jugendlicher Computerfans, die sich weltweit vernetzt haben, Software austauschen und bearbeiten. Sie kreieren eigene Sounds, eigene Grafiken und eigene Kommunikationsformen.

Computerspiele als fixer Bestandteil der jugendlichen Konsum- und Erlebniswelt machen Spaß, schulen die Konzentration, fördern Geschicklichkeit und Reaktionsfähigkeit. Viele von ihnen gehen aber von gewaltverherrlichenden und menschenverachtenden Konzepten aus. Der Wettbewerb COMPUTER UND SPIELE sollte in der Schule die Auseinandersetzung mit diesen Aspekten von Computerspielen anregen und intensivieren.

SchülerInnen aller Schulstufen wurden eingeladen, gemeinsam mit ihren LehrerInnen Ideen zu Computerspielen zu entwerfen und/oder Spiele speziell für das Medium Computer zu entwickeln.

Die Einreichung war in drei Kategorien möglich:
1) Konzept: Spielbeschreibung, graphische Entwürfe, Skizzen etc.
2) Demo: Spielbeschreibung und ausprogrammierte Spielsequenzen
3) Spiel: Spielbeschreibung und vollspielbare Version

Zugelassen waren jede gängige Software sowie Programmierhilfsmittel (Tools).

Eine unabhängige Fachjury unter Vorsitz von Projektleiter Dr. Seppo Gründler (Institut für elektronische Musik der MHS Graz) wählte unter Berücksichtigung der unterschiedlichen Altersstufen der TeilnehmerInnen aus den Einsendungen jene Arbeiten aus, die auf der Ars Electronica 1994 präsentiert werden.

Die GewinnerInnen nehmen im Rahmen von Ars Electronica 1994 an einem dreitägigen Computercamp unter der Leitung des TOP JOB Teams teil.

Die Abschlußpräsentation findet am 24.Juni 1994 um 14.30 Uhr im Stiftersaal statt.

Nähere Informationen: ÖKS – 1070 Wien, Stiftgasse 6
Tel.: 0222/523 57 81/DW18 (Dr. Sirikit M. Amann)

WALTRAUD BARTON
(ÖKS-Öffentlichkeitsarbeit)

Results of a National Competition Organized by the Austrian Cultural Service in Cooperation with the Federal Ministry of Education and Art and Ars Electronica

Dealing with computer technologies is a matter of course for young people. There is already a "demo-group" scene of internationally networked young computer fans. They exchange and process software and create their own sounds, graphics and forms of communication.

Computer games, as a permanent part of young peoples' world of consumption and experience are fun, train the concentration, challenge the player's dexterity and ability to react. However, many of them are based on misanthropic concepts which glorify violence.

The COMPUTER AND GAMES contest was intended to stimulate and intensify the discussion of these aspects of computer games.

Students of all ages were invited to develop their ideas relating to computers and/or design games especially for the computer together with their teachers.

Submission was possible in three categories:
1) Concept: description of the game, graphic designs, sketches, etc.
2) Demo: description of the game and programmed game sequences
3) Game: description of the game and fully playable version

All common software and programming tools were allowed. An independent jury of experts under the direction of project head Dr. Seppo Gründler (Institute of Electronic Music at the College of Music, Graz) selected from the submissions those works which will be presented at the Ars Electronica 94. The participants' ages were also taken into consideration. The winners will participate in a three-day computer camp directed by the TOP JOB team as part of Ars Electronica 94. The final presentation will take place at 2:30 PM on June 24, 1994 in the Stiftersaal.

For more information: ÖKS, Stiftgasse 6, 1070 Vienna;
Tel.: 0222/523 57 81, ext. 18 (Dr. Sirikit M. Amann)

WALTRAUD BARTON
(ÖKS Public Relations Department)

MIT DEN AUGEN DER ARCHITEKTUR
with the eyes of architecture

 OFFENES KULTURHAUS

Ein Ausstellungsprojekt
an exhibition project

The power of architecture has waned. The modern age is no longer able to ask the central question; the postmodern age has never really taken the risk.
This project has assumed the task of introducing young architects who are attempting (outside the mainstream of an instrumentalized purposive rationalism) to solve architectural problems without subjecting themselves to the actual implementation of architectural conventions, without replacing the concrete public sphere with an anonymous, hypothetical one. When the decision to proceed in this way is made, one chooses the difficult, complex path which avoids assembly-line production and projects made for the drawer. This is perhaps the only way to recognize architecture within its contradictory and intricate content.

Die Kraft der Architektur ist erlahmt. Die Moderne ist nicht mehr in der Lage, die zentralen Fragen zu stellen; die Postmoderne hat es nie wirklich riskiert.
Das Projekt stellt sich die Aufgabe, junge ArchitektInnen vorzustellen, die abseits des Mainstreams eines instrumentalisierten Zweckrationalismus versuchen, architektonische Aufgaben zu lösen, ohne die reale Umsetzung architektonischen Konventionen zu unterwerfen, ohne die konkrete Öffentlichkeit durch eine anonyme, hypothetische zu ersetzen. Fällt die Entscheidung für solch eine Vorgangsweise, wählt man den schwierigen, komplexen Weg, abseits von Fließbandproduktion und Schubladenprojekten. Vielleicht ist dies jedoch der einzige, der Architektur in ihrem widersprüchlich-vielschichtigen Gehalt erkennt.

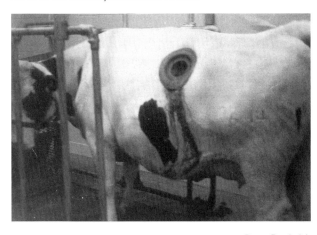

„geschenktes Photo" (1992/93 von Tim Ventimiglia an Wolfgang Tschapeller)

„Photograph Given as a Present" (1992/93 from Tim Ventimiglia to Wolfgang Tschapeller)

This project is intended to reflect on fundamental issues of architecture in a sensorial way by using unconventional perspectives. Shifted points of view illuminate society's relationship to aesthetic judgement. Architecture attempts to direct the attention towards the opposite side as a means of transportation, as an opportunity, and formulate replies.
Five artists/architects demonstrate the access doors leading to changed view-points of architecture and its principles. The important issue is not architecture as a constructed ambience, but the structuring of the architectural eye in a medialized world.

Das Projekt will durch unkonventionelle Perspektiven Grundfragen des Architektonischen auf sinnliche Weise reflektieren. Verschobene Blickwinkel durchleuchten das Verhältnis der Gesellschaft zum ästhetischen Befinden. Architektur versucht sich als Transportmittel, als Möglichkeit, die Aufmerksamkeit auf die Gegenseite zu richten und Entgegnungen zu formulieren.
Fünf Künstler-ArchitektInnen zeigen Einstiegsluken für veränderte Sichtweisen auf Architektur und ihre Prinzipien. Es geht nicht um Architektur als gebautes Ambiente, sondern um die Strukturierung des architektonischen Blicks in einer medialisierten Welt.

TeilnehmerInnen:

ADOLPH-HERBERT KELZ / BRIGITTE LÖCKER

So wie die Architektur immer auf die Verkörperung des Denkens hinweist, verfolgt das Projekt „Denklaboratorium" die Absicht, ein Ereignisfeld zu schaffen, das eine Welt projizieren im Stande ist, in der Art eines simultanen Ereigniskontinuums, eines totalen, reflexiven Wirkungsgefüges, das in seiner gesamten Handlungsdichte sowie seiner architektonischen Morphologie mit Konzepten aus Literatur und Film assoziativ wie auch konkret in Analogie steht.

PRINZGAU / PODGORSCHEK

x-house or shrine for the goddess of inbetween, Chicago 1993

If art is an organism, like a kind of hermaphrodite, then one kind of art or the other is the ear, the testicle, the cerebellum, the armpit hair, the blood, the excrement, the melanin, the eye, the liver, the rage ... (pod 91)

Participants:

As architecture has always called attention to the embodiment of thought, the "Denklaboratorium" ("Think Laboratory") project intends to create a field of events capable of projecting a world in a manner similar to a continuum of events, a total reflective structure of influences which is associatively and concretely analogous to concepts from literature or film in its total density of action and in its architectural morphology.

"x-house or shrine for the goddess of in between", Chicago 1993

If art is an organism, like a kind of hermaphrodite, then one kind of art or the other is the ear, the testicle, the cerebellum, the armpit hair, the blood, the excrement, the melanin, the eye, the liver, the rage... (pod 91)

WOLFGANG TSCHAPELLER

„Gewinn von Tiefenschärfe bedeutet Verlust von Lichtintensität. Das Objektiv (1) ist immer eine Übereinkunft, eine Art Vertrag des abwechselnden Schweigens zwischen eintretender Lichtstärke und Sehschärfe. Versuche ich sehr scharf zu sehen, wird es dunkel, und zwar in einem Maß, daß das Sehen selbst obskur wird (2) und das zu Betrachtende nicht mehr sichtbar sein könnte.

Der blinde Fleck umschreibt die Gegend im Auge, die keine visuelle Informationen liefert. Der bewegte Blick läßt den blinden Fleck als Zensurbalken (3) (reflexiv) über die Umgebung gleiten. „Vom Auge gehen unsichtbare

"A gain in depth of focus means a loss of illumination. The objective (1) is always an agreement, a kind of contract of alternating silence between the ensuing illumination and sharpness of vision. If I attempt to see very clearly, everything becomes dark, and precisely to the extent that seeing itself becomes obscure (2) and that which is to be looked at is no longer visible.

"The blind spot refers to the area in the eye which does not deliver visual information. The roving glance allows the blind spot to sweep

across the environment as a censor's stroke (3) (reflective). 'Invisible rays which scan the environment are emitted from the eye. These rays look like needles (cf.: knitting needles) or legs. Whatever they touch disappears. The body's legs are needles. Seeing is deleting. What is read is eaten (read is deleted). The grounds are white. What is read is white.'"
(1) see collective court of arbitration, (2) darkens, (3) sleeping place (reflective)

Strahlen aus, die die Umwelt abtasten. Diese Strahlen sehen wie Nadeln (vgl.: Stricknadeln) oder wie lange Füße aus. Was immer sie berühren verschwindet. Die Füße des Körpers sind Nadeln. Sehen ist Löschen. What is read is eaten (gelesen ist gelöscht). The grounds are white. What is read is white."
(1) kollektives Schiedsgericht, (2) sich verdunkelt, (3) Schlafstelle (reflexiv)

Traum
Dream

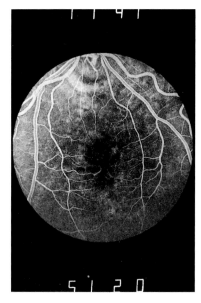

Auge
Eye

HANS PETER WÖRNDL

Guglhupf – geöffnet – von Seeseite gesehen; Sperrholz, Holz, Aluminium, Glas, Siebdruck, Mondsee, 1993

„Guglhupf – opened – seen from the side facing the lake"; plywood, wood, aluminum, glass, silkscreen print; Mondsee, 1993

Harry, Ilse und Viktor stellen in der Galerie Maerz aus.

Public Intervention
Galerie Maerz, Landstraße 7
8. – 29. Juni, Montag – Freitag, 15.00 – 18.00 Uhr, Samstag, 10.00 – 13.00 Uhr

BIOGRAPHIEN

FAREED ARMALY
was born in Iowa. After completing high school, he moved between various states, universities and art academies. In 1988 and 1989 he edited and published two journals on music and culture (Terminal Zone, R.O.O.M.). He is an artist.

PETER ASSMANN
was born in 1963. He studied art history, history and German. Since 1992 director of the Stifterhaus gallery as wel as the regional gallery in the O.Ö. Landesmuseum in Linz. Since 1993 curator for the art history and graphic collections of the O.Ö. Landesmuseum Linz

ANNA BICKLER (SUPREME PARTICLES)
Born 1964 in Friedberg. Studied film at th Academy of Design, Offenbach; since 1991 on the Institute for New Media, Städelshule, Frankfurt. Participant in various festivals including the 4. Eureopean Media Art Festival, Osnabrück; Olympic Museum, Sarajevo; Polytechnichs Central, London; Videonale, Bonn; and the German Video Awards: The 50 Best.

MICHAEL BIELICKY
Born 1954 in Prague. He studied medicine at Düsseldorf University and recieved his MA in 1987 at the Acadamy of Fine Art in Düsseldorf from Nam June Paik. In 1991 he founded the Video Department at the Academy of Fine Arts in Prague, where he is teaching as Associate Professor. His work has been shown at MOMA, NYC, Documenta 8, Kassel and numerous other shows around the world.

LOREN CARPENTER
is the inventor and President of CINEMATRIX™ Interactive Entertainment Systems Inc. He is a pioneer in the field of computer graphics and, together with his wife Rachel, founded Cinematrix to explore the intersection of computers and art. Cinematrix is currently focusing on the development of interactive audience participation technology.

STEPHEN GALLOWAY
American, born in Pennsylvania, he came to Ballet Frankfurt in 1985.

ALBINA MIRELLA D'URBANO
Born 1955 in Tivoli. Studied painting under Prof. Enzo Brunori at the Accademia de Bella Arti, and visual communication on the Academy of Arts in Berlin. Grant Recipient in Berlin, Darmstadt and Paris. Since 1990 artistic work at the Institute for New Media in Frankfurt. Participant in group and solo exhibitions in Berlin, Genoa, Frankfurt, Rom, Mannheim, Venice (Biennial) and Milan.

CHRISTOPH FÜRST
Born 1965 in Freistadt. Since 1987 has studied with Prof. Gsöllpointner at the Accademy for Artistic and Industrial Design in Linz (Master class for Metal and Object Design). 1992/93 recieved a grant to study at the Art Academy, Düsseldorf. Works in metal sculpture, public sculpture, architectural conception, as well as product and furniture design.

KARL GERBEL
Born 1939 in Linz. Since 1984 chairman of the board of directors for the Linzer Veranstaltungsgesellschaft (LIVA) and is responsible for activities at the town hall, sport center, Brucknerhaus, stadium, Posthof and the childrens cultural center. He is also a member of the board of directors for ARS ELTRONICA.

FAREED ARMALY
ist in Iowa geboren. Nach dem Abschluß des Gymnasiums besuchte er Universitäten und Kunstakademien in verschiedenen Ländern. 1988 und 1989 gab er zwei Journale heraus, die sich mit Musik und Kultur beschäftigten (Terminal Zone, R.O.O.M.). Er ist Künstler.

PETER ASSMANN
Mag., Dr. 1963 geboren. Studium der Kunstgeschichte sowie Geschichte und Germanistik. Seit 1992 Leiter der Galerie im Stifterhaus sowie der Landesgalerie am O.Ö. Landesmuseum, Linz. Seit 1993 Kustos für Kunstgeschichte und für die Graphische Sammlung des O.Ö. Landesmuseums.

ANNA BICKLER (SUPREME PARTICLES)
1964 in Friedberg geboren. Filmstudium an der Hochschule für Gestaltung, Offenbach; seit 1991 am Institut für Neue Medien, Städelschule Frankfurt. Teilnahme an verschiedenen Festivals, u.a. 4. European Media Art Festival, Osnabrück; Olympisches Museum, Sarajewo; Polytechnics Central London; Videonale Bonn; Deutscher Videopreis: Die 50 Besten.

MICHAEL BIELICKY
1954 in Prag geboren. Studium der Medizin und an der Akademie der bildenden Künste in Düsseldorf, wo er 1987 bei Nam June Paik abschloß. 1991 gründetete er eine Abteilung für Video an der Akademie Prag, wo er als Gastprofessor tätig ist. Seine Arbeiten wurden weltweit ausgestellt, u.a. im MOMA New York und auf der documenta 8 in Kassel.

LOREN CARPENTER
Erfinder und Präsident von CINEMATRIX™ Interactive Entertainment Systems, Inc. Pionier auf dem Gebiet der Computergrafik. Loren Carpenter und seine Frau Rachel gründeten Cinematrix zur Erforschung der Schnittpunkte von Computer und Kunst. Zur Zeit beschäftigt sich Cinematrix vor allem mit der Entwicklung einer Technologie für interaktive Publikumspartizipation.

STEPHEN GALLOWAY
Amerikaner, geboren in Pennsylvania, kam 1985 zum Ballett Frankfurt.

ALBINA MIRELLA D'URBANO
1955 in Tivoli bei Rom geboren. Studium der Malerei bei Prof. Enzo Brunori an der Accademia de Bella Arti, der visuellen Kommunikation an der Hochschule der Künste in Berlin. Stipendiate in Berlin, Darmstadt und Paris. Seit 1990 künstlerische Tätigkeit am Institut für Neue Medien in Frankfurt. Einzelausstellungen und Ausstellungsbeteiligungen u.a. in Berlin, Genua, Frankfurt, Rom und Mannheim, Venedig (Biennale), Mailand.

CHRISTOPH FÜRST
1965 in Freistadt geboren. Seit 1987 bei Prof. Gsöllpointner Studium an der Hochschule für künstlerische und industrielle Gestaltung in Linz (Meisterklasse für Metall und Objektgestaltung. 1992/93 Stipendiat an der Kunstakademie Düsseldorf. Beschäftigung mit Metallplastik, Großplastik im öffentlichen Raum, architektonischer Konzeption, sowie Produkt- und Möbeldesign.

KARL GERBEL
1939 in Linz geboren. Seit 1984 Vorstandsdirektor der Linzer Veranstaltungsgesellschaft (LIVA) mit Brucknerhaus, Stadt- und Sporthalle, Stadion, Posthof und Kinderkulturzentrum. Mitglied des Direktoriums von Ars Electronica.

FRIEDRICH GULDA

geboren 1930 in Wien und dort eine Wunderkind-Karriere. Hat sich, nachdem er alle Möglichkeiten einer Virtousenlaufbahn ausgekostet hatte, zum verwegenen Aussteiger und Außenseiter des Musikbetriebes entwickelt. Wo er erscheint, begibt sich Musizieren außerhalb der zum Fetisch erhobenen Norm: Gulda spielt nach Ansage; Gulda protesiert vornehmlich in eigenen Kompositionen, vornehmlich in eigenen Kompositionen gegen die hochmütige Grenzziehung zwischen E- und U-Musik; Gulda gibt dem Jazz in seinen Programmen den gebührenden Stellenwert; Gulda ist die treibende Kraft der Bemühungen, die starren Rituale des Konzertlebens aufzubrechen; Gulda führt einen zähen Kampf für die Wiederentdeckung und Anerkennung des improvisatorischen Elements in der Musik; Gulda wünscht sich das Musizieren lebendiger, farbiger, weniger verbissen.

FRANZ HANNES

1960 in Innsbruck geboren. Studium an den Akademien der bildenden Künste in München und Wien (Bildhauerei und Malerei). 1990 Assistenz bei Leiko Ikerrura, Sommerakademie Salzburg. Lebt in Wien. Ausstellungen u.a. in Graz, Innsbruck, Paris, Laibach, Berlin.

KARIN HAZELWANDER

geb.in Österreich, lebt in Wien. Ausstellungen im Museum Moderner Kunst, Wien (1984). Museum Ludwig, Köln (1988). Pac Gallerie, Chicago (1989). Museum des 21.Jahrhunderts, (1989). Galerie Zacheta, Warschau (1991). Frontiera, Bozen (1992). Museum in Progress/Reise zu den Quellen, (1993). O.Ö. Landesmuseum, Linz (1994). Steirischer Herbst, Graz (1993 und 1994).

PERRY HOBERMAN

präsentierte Installationen, Spiele, Skulpturen und Performances in den USA und Europa. Hoberman wird von Postmasters Gallery vertreten. Zur Zeit unterrichtet er an der Cooper Union School of Art.

KATHY RAE HUFFMAN

ist freischaffende Kuratorin und publiziert über Medien und zeitgenössische Kunst. Kathy Rae Huffmann war Kuratorin und Produzentin für den Contemporary Art Television Fund. 1987 ko-kuratierte sie The Arts for Television. Seit 1991 unterrichtet sie über Kunstvideos, interaktives Fernsehen und die Geschichte von Künstlern und Fernsehen.

GOTTFRIED HÜNGSBERG

1944 in Bayern geboren. Betreibt ein Ingenieurbüro in München.

ELFRIEDE JELINEK

1946 in Mürzzuschlag geboren. Studium der Theaterwissenschaft, Kunstgeschichte und Musik in Wien. Lebt als freie Schriftstellerin in Wien, München und Paris. Publikationen (Auswahl): Die Liebhaberinnen, 1975. Die Ausgesperrten, 1980. Die Klavierspielerin, 1983. Lust, 1989. Wolken.Heim., 1990. Malina. Ein Filmbuch, 1991 (Drehbuch zu Werner Schroeters Verfilmung von Ingeborg Bachmanns „Malina").

ROBERT JELINEK

1970 in Pilsen (CR) geboren. Studium an der Hochschule für industrielle und künstlerische Gestaltung Linz (Malerei und Graphik, visuelle Gestaltung), an der Kunstakademie Düsseldorf (freie Kunst) und an der Akademie für bildenden Künste Wien (Malerei und Graphik). SABOTAGE I – XVII a unter anderem in Wien, Kassel, Köln, Salzburg, Prag, New York, Chicago.

STEFAN KARP (SUPREME PARTICLES)

1963 in Frankfurt /Main geboren. Studium der Philosophie und Produktgestaltung. Gründungsmitglied der Gruppe "Interfacelifting", Entwicklung und Ausarbeitung von Konzepten im Bereich Benutzeroberflächen (z.B. Flugticketautomat), Multimedia und terminaler Hardware. Austellung auf der Buchmesse ´93 im Bereich electronic publishing mit der Gruppe interfacelifting.

FRIEDRICH GULDA

was born in 1930. A child prodigy, he went on to explore all the avenues of a virtuoso career, before developing into an audacious musical non-conformist. Wherever he apperrs on the scene, music-making loses its conventional aura. Gulda announces his programmes; he protests – mainly in his own compositions – aigainst the boundaries arrogantly drawn between "light" and "serious" music; he gives jazz its due place in his programmes; he is the driving force behind the efforts to break up the rigid ritual of concert routine; he carries on a tough struggle for the rediscovery and recognition of improvisation in music; he would like to see music become more alive and more colorful, with less grim determination.

FRANZ HANNES

Born 1960 in Innsbruck. Studied at the Academy of Applied Arts in Munich and Vienna (sculpture and painting). 1990 assistant to Leiko Ikerrura at the Summer Academy, Salzburg. Lives in Vienna and has had exhibitions in Graz, Innsbruck, Paris, Laibach, and Berlin.

KARIN HAZELWANDER

was born in Austria and lives in Vienna. Exhibitions in the Museum of Modern Art, Vienna (1984); Museum Ludwig, Cologne (1988); Pac Gallery, Chicago (1989); Museum of the 21st Century (1989); Gallery Zacheta, Warsaw (1991); Frontiera, Bozen (1992); Museum in Progress/Reise zu den Quellen (1993); O.Ö. Landesmuseum, Linz (1994); Steirischer Herbst, Graz (1993 and 1994).

PERRY HOBERMAN

has presented installations, spectacles, sculptures and performances throughout the United States and Europe. Hoberman is represented by Postmasters Gallery, and currently teaches at the Cooper Union School of Art.

KATHY RAE HUFFMAN

is a freelance curator and writer on media and contemporary art. Ms. Huffman was Curator/Producer of the Contemporary Art Television Fund. In 1987 she co-curated The Arts for Television. Since 1991 Ms. Huffman has lectured extensively on art videos, interactive TV, and the history of artists and television.

GOTTFRIED HÜNGSBERG

Born 1944 in Bavaria. Runs an engineering office in Munich.

ELFRIEDE JELINEK

Born 1946 in Mürzzuschlag. Studied theater arts, art history and music in Vienna. Lived as a freelance writer in Vienna, Munich, and Paris. A selection of her works include: Die Liebhaberinnen, 1975; Die Ausgesperrten, 1980; Die Klavierspielerin, 1983; Lust, 1989; Wolken. Heim.,1990; Malina. Ein Filmbuch, 1991 (a screenplay for Werner Schroeters filming of Ingeborg Bachmans "Malina").

ROBERT JELINEK

Born 1970 in Pilsen (CR). Studied at the Academy of Industrial and Artistic Design in Linz (painting and graphic arts, visual design), the Art Academy Düsseldorf (free art) and at the Academy of Fine Arts in Vienna (painting and graphic arts). SABOTAGE I-XVI in Vienna, Kassel, Cologne, Salzburg, Prague, New York and Chicago.

STEFAN KARP (SUPREME PARTICLES)

Born 1962 in Frankfurt. Studied philosophy and product design. Founding member of the group INTERFACELIFTING, worked on development and preparation of concepts for user interfaces (an automatic teller for airplane tickets), multimedia, and terminal hardware. Electronic publishing exhibition at the '93 Book Convention with the group INTERFACELIFTING.

ADOLPH-HERBERT KELZ
studied at the TU in Graz and the Summer Academy at Abraham. Projects: Contest and Realization University Graz/ Heinrichstraße (with Kapfhammer/ Wegan/Kossdorf); Housing development Ziegelstraße/Graz (with Soran). Exhibitions: Europalia; Architecture from Graz/ Brussels; Visionary Architecture/Vienna; L´Europe des Createurs-Les Utopies 89 Grand Palaise/Paris; Foundation of RHIZOM-Architects; the Steiermark Prize for Architecture; an office in Graz.

MICHAEL KLEIN
born in 1960 in Wuppertal. Studied physics and philosophy in Wuppertal and Tübingen. Assistent at the Institute for Physical and Theoretical Chemistry at the University of Tübingen, scientific and artistic assistent at the Städelschule Institute for New Media with Peter Weibel. He is co-author of the book "A Chaotic Hierarchy", W.S. Singapore 1991, and founded an interdisciplinary research group Engadyn in 1990.

MARTIN KUSCH
Born 1964 in Leoben. Studied visual media design under Peter Weibel in Vienna. He has participated in video festivals in Austria, Poland, Turkey, Germany, and England.

JARON LANIER
is best known as a pioneer of virtual reality, a phrase he coined. The co-inventor of many fundamental VR technologies such as interface gloves and VR networking, and an innovator in the field of visual programming. Jaron is also an accomplished composer, visual artist and author. He has performed with Philip Glass, Ornette Coleman, Terry Riley, Barbara Higbie and Stanley Jordan.

BRIGITTE LÖCKER
studied at the TU in Graz and Cooper Union in New York; worked with Raimund Abraham and John Hejduk; participated in an internship program with Dore Ashton; received practical training from Allen McCollum and Kim Wang in New York; doctoral work with Karin Wilhelm in Graz. Projects: exhibitions in Graz, Vienna, New York, Hungary, and Houghton.

HERMEN MAAT
born 1963 in the Netherlands. Studied at the Rietveld Academy, Amsterdam. Participant postgraduate program Van Eyck Academy, Maastricht. Most of his works concern themselves with identity in the broadest sense. Exhibitions in The Netherlands, Danmark and USA. Received several grants and scholarships.

GIDEON MAY (SUPREME PARTICLES)
Born 1964 in Rotterdam. He has worked as a fashion photography assistant. Until 1986 he worked as a special effects artist for film and TV as well as working three years as an assistant cameraman. In 1987 he began work programming computer graphics for various projects, including The Table of Spirits for Ars Electronica 1993 and work for the Softimage Company.

RON MILTENBURG
Born 1951 in Rotterdam. Studied eastern philosophy in India, Switzerland, and the USA; and has studied art history in Amsterdam. Researches the various paradigms of metonyms and metaphors. He is a freelance publicist, curator, television and video producer. Exhibitions in the USA and the Netherlands.

PAUL MODLER (SUPREME PARTICLES)
Born 1959 in Aschaffenburg. Composer and musician in the conflicting fields of improvised and notated music, acoustic and electronic sound generation. Developed an algorithmic sound system.

ADOLPH-HERBERT KELZ
Studium an der TU Graz, Sommerakademie bei Raimund Abraham. Projekte: Wettbewerb und Realisierung Uni Graz/Heinrichstraße (mit Kapfhammer/Wegan/Kossdorff), Wohnanlage Ziegelstraße/Graz (mit Soran) Ausstellungen: Europalia, Architektur aus Graz/Brüssel, Visionäre Architektur/Wien, L´Europe des Createurs–Les Utopies 89 , Grand Palais/Paris, Gründung von RHIZOM-Architekten, Architekturpreis des Landes Steiermark, Büro in Graz.

MICHAEL KLEIN
1960 in Wuppertal geboren. Studium der Physik und Philosophie in Wuppertal und Tübingen. Assistent am Institut für Physikalische und Theoretische Chemie der Universität Tübingen, wissenschaftlich-künstlerischer Mitarbeiter am Städelschule- Institut für Neue Medien bei Peter Weibel. Koautor des Buches „A Chaotic Hierarchy", W.S. Singapore 1991, und Gründer der interdisziplinäre Forschungsgruppe Engadyn.

MARTIN KUSCH
1964 in Leoben geboren. Studium der visuellen Mediengestaltung bei Peter Weibel in Wien. Teilnahme an Videofestivals u.a. in Österreich, Polen, Türkei, Deutschland und England.

JARON LANIER
wurde als Pionier der Virtuellen Realität bekannt, ein Begriff übrigens, den er prägte. Er ist Miterfinder vieler grundlegender VR-Technologien wie dem Interface Handschuh oder von Virtuell-Reality-Netzwerken. Er ist ein Neuerer auf dem Gebiet des visuellen Programmierens. Überdies ist er als Komponist, visueller Artist und Autor tätig. Er trat mit Philip Glass, Ornette Coleman, Terry Riley, Barbara Higbie und Stanley Jordan auf.

BRIGITTE LÖCKER
Studium an der TU Graz und Cooper Union/New York, Mitarbeit bei Raimund Abraham und John Hejduk, Internship-Programm bei Dore Ashton, Practical-Training bei Allen McCollum und Kim Wang / New York, Doktoratsstudium bei Karin Wilhelm/Graz. Projekte: Ausstellungen in Graz, Wien, New York, Ungarn, Houghton.

HERMEN MAAT
1963 in Holland geboren. Studium an der Rietveld Academy in Amsterdam. Postgraduatestudium an der Van Eyck Academy in Maastricht. Die meisten seiner Arbeiten beschäftigen sich mit dem Thema Identität im weitesten Sinne. Ausstellungen in Holland, Dänemark und den USA. Erhielt verschiedene Stipendien.

GIDEON MAY (SUPREME PARTICLES)
1964 in Amsterdam geboren. Programmiert seit 1977; Arbeit als Modephotographie-Assistent. Anschließend zwei Jahre Trickaufnahmen für Film und Fernsehen sowie drei Jahre Kameraassistenz. Seit 1987 arbeitet er als Computergrafik-Programmierer für verschiedene Projekte, u.a. The table of spirits, Ars Electronica 1993; Work for the Softimage Company.

RON MILTENBURG
1951 in Rotterdam geboren. Studium der östlichen Philosophie in Indien, der Schweiz und den USA. Studium der Kunstgeschichte in Amsterdam. Erforscht die verschiedenen Paradigmen von Metonymien und Metaphern. Freischaffender Publizist, Kurator. TV- und Videoarbeiten. Ausstellungen in den USA und Holland.

PAUL MODLER (SUPREME PARTICLES)
1959 in Aschaffenburg geboren. Lebt seit 1987 in Berlin. Komponist und Musiker im Spannungsfeld improvisierter und notierter Musik, akustischer und elektronischer Klangerzeugung. Entwickelt algorithmische Musiksysteme.

CHRISTIAN MÖLLER

1959 in Frankfurt geboren. Studium der Architektur in Frankfurt, anschließend Stipendiat bei Gustav Peichl an der Akademie der Bildenden Künste in Wien. Von 1988 bis 1990 arbeitete er im Architekturbüro Behnisch und Partner in Stuttgart und leitete seit 1991 ein eigenes Büro in Frankfurt. Dort hatte er an der Städelschule an der Architekturklasse einen Lehrauftrag für virtuelle Architektur.

MONIQUE MULDER

1965 in Rotterdam geboren. Studium an der Kunstakademie, Den Haag (Architektur/Interior Design); Postgraduate-Ausbildung (Computergraphik und Animation). Regieassistenz und TV-Produktionsassistenz; Produktionsleiterin und Stylistin für Modephotographie; Stylistin für Architektur und Interior Design; Produzentin für 3-D-Animationsprojekte.

PONTON – VAN GOGH TV

Seit 1986 verschiedene Radio- und Fernsehprojekte, z.B. interaktive TV-Projekte bei Ars Electronica 1986, 1989, 1990. 1986 Gründung von Ponton European Media Art Lab in Hamburg. 1992 Piazza virtuale, 100 Tage interaktives Live-Fernsehen bei documenta 9 in Kassel. 1993 Piazza virtuale live in Japan.

PRINZGAU/podgorschek

geboren in Österreich bzw. Yugoslawien. Arbeitsschwerpunkte: öffentlicher Raum, Objektarchitekturen, Film, Komplementär- und Infiltrationsarchitektur, Auszug von Ausstellungen, Filme/Festivals: Musée des arts d´Afrique et Océanie/Paris, Warehouse Loft/Chicago, Festival Rotterdam, Cannes / Wettbewerb, Triennale Mailand.

GÜNTHER RABL

1953 in Linz geboren. Matura an der HTL (Elektrotechnik). Seit der Gründung eines eigenen Experimental-Studios Kompositionen. 1979 und 1989 Staatsstipendium für Komposition. Entwickelte ein Software-Paket zur numerischen Musikverarbeitung auf dem PC. Lehrtätigkeit am Institut für Elektroakustik an der Musikhochschule Wien.

CONSTANZE RUHM

1965 in Wien geboren. 1985 Beginn des Studiums an der Hochschule für angewandte Kunst, Meisterklasse für visuelle Mediengestaltung. Seit 1991 studiert sie am Institut für Neue Medien an der Städelschule Frankfurt/M. Einzelausstellungen in Graz und gemeinsam mit Peter Sandbichler in Wien. Ausstellungsbeteiligungen in Wien, Paris, Istanbul, Frankfurt, Hamburg, Graz, New York. 1993 Prix Ars Electronica: Anerkennung für Computergraphik.

PETER SANDBICHLER

1964 in Kufstein geboren. Bis 1983 Fachschule für Holz- und Steinbildhauerei. Anschließend Studium bei Prof. Martin Knox an der Arts Students League, NYC; in der Meisterklasse Wander Bertoni an der Hochschule für angewandte Kunst, Wien und bei Prof. Bruno Gironcolli an der Akademie der Bildenden Künste in Wien. Einzelausstellungen in den Galerien Krinzinger und Grita Insam. Ausstellungsbeteiligungen u.a. in Graz, Wien, Montreal, Bozen, Los Angeles, Paris, Berlin, Rostock, Barcelona, Linz.

MICHAEL SAUP (SUPREME PARTICLES)

1961 in Jungingen geboren. Studium der Musik am Dominican College San Rafael, Kalifornien, USA; der Informatik an der FH Furtwangen und der Visuellen Kommunikation an der Hochschule für Gestaltung Offenbach. Lehrtätigkeit für Video und Computer an der Akademie der bildenden Künste, München und an der Hochschule für Gestaltung Offenbach. Seit 1990 künstlerisch-wissenschaftlicher Mitarbeiter am Institut für Neue Medien, Städelschule Frankfurt. 1993 Gastprofessur an der Kunsthochschule Bremen.

CHRISTIAN MÖLLER

Born 1959 in Frankfurt. Studied architecture in Frankfurt. He then recieved a grant to study at the Academy of Fine Arts in Vienna under Gustav Peichel. Between 1988 and 1990 he worked at the architectural agency Behnisch and Partner in Stuttgart. Since 1991 he has run his own agency in Frankfurt where he has also taught classes in virtual architecture for the department of architecture at the Städelschule.

MONIQUE MULDER

Born 1965 in Rotterdam. Studied at the Art Academy, Den Haag (architecture, interior design); and recieved postgraduate training in computer graffics and animation. She has worked as an assistant director, production assistant for television, production director and stylist for fashion photography, stylist for architecture and interior design, and as a producer for a 3-D animation.

PONTON – VAN GOGH TV

Since 1986 various radio and TV projects, e.g. interactive TV projects at Ars Electronica 1986, 1989, 1990. In 1986 they founded Ponton European Media Art Lab in Hamburg. In 1992 Piazza virtuale, 100 days of inter-active live TV at documenta 9 in Kassel. In 1993 Piazza virtuale live in Japan.

PRINZGAU/podgorschek

Born in Austria and Jugoslavia. They focus their efforts on: public spaces; the architecture of complimentation, infiltration and objects; film; and the arrangement of exhibitions. Films/Festivals: Museé des arts d'afrique et oceanie, Paris; Warehouse loft, Chicago; Festival Rotterdam; Cannes; Triennale Milan.

GÜNTHER RABL

Born 1953 in Linz. Completed studies in electrical engineering. He has been composing since the opening of his own experimental studio. In 1979 and again in 1989 he recieved the state grant for composition. He has developed his own software package for the numerical manipulation of music on the PC and is teaching at the Institute for Electro-Acoustics at the Music Academy in Vienna.

CONSTANZE RUHM

Born 1965 in Vienna. 1985 began studying at the Masterclass for Visual Media Design, Academy of Applied Arts, Vienna. In 1991 she began studies at the Institute for New Media, Städelschule, Frankfurt. She has had a solo exhibition in Graz and a shared exhibition with Peter Sandbichler in Vienna. Ms. Ruhm has also taken part in group exhibitions in Vienna, Paris, Istanbul, Frankfurt, Hamburg, Graz and New York. In 1993 she recieved Honorable Mention for computer graphics at the 1993 Prix Ars Electronica.

PETER SANDBICHLER

Born 1964 in Kufstein. Studied wood and stone sculpture until 1983. He has also studied under Prof. Martin Knox at the Arts Student League, NYC; under Wander Bertoni at the Academy of Applied Arts, Vienna; and under Prof. Bruno Gironcolli at the Academy of Fine Arts, Vienna. He has had solo exhibitions at Gallery Krinzinger and Grita Insam, and has taken part in group exhibitions in Graz, Wien, Montreal, Bozen, Los Angeles, Paris, Berlin, Rostock, Barcelona and Linz.

MICHAEL SAUP (SUPREME PARTICLES)

Born 1961 in Jungingen. Studied music at Dominican College, San Rafael, California; computer science in Furtwangen and visual communications at the Academy of Design in Offenbach. He has been teaching video and computer at the Academy of Fine Arts in Munich and the Academy of Design in Offenbach. He has also been the artistic reasearch assistant at the Institute for New Media, Städelschule in Frankfurt since 1990. In 1993 he took up the role of Guest Professor at the Art Academy, Bremen.

JOACHIM SAUTER (ART+COM)
Born 1959. Studied at the Academy of Arts and the German Film and Television Acadamy, Berlin. He has worked, taught, published, exhibited and recieved international awards. (e.g. the prize for interactive art at Ars Electronica '92 for his work "Zerseher"). Founding member of Art + Com, Berlin. He is now Professor of Design for Digital Media, Academy of Arts.

ELLIOTT SHARP
Composer/multi-intrumentalist Elliott Sharp leads the groups CARBON and ORCHESTRA CARBON, DYNERS CLUB, BOOTSTRAPPERS, and TERRAPLANE. Sharp records extensively for a number of independent labels in the United States, Canada, and Germany with recent releases of CARBON, ORCHESTRA CARBON, and 'DYNERS CLUB. DIGITAL appears on the new disc by Kronos Quartet. In 1977, he founded the zOaR label to release his own and other extreme musics.

JEFFREY SHAW
Born 1944 in Melbourne. Presently director of the ZKM Institute for Visual Media, Karlsruhe. Selected works: Corpocinema, "Sigma Projects", Museumsplein, Amsterdam 1967; Diadrama, Lantaren Theater, Rotterdam 1974; Viewpoint, 6th Biennale de Paris, Musée d'Art Moderné, Paris 1975; Eve, "Multimedia 3", Center for Art and Media (ZKM) Karlsruhe.

SOLDIER STRING QUARTET
Since their founding in 1985, the Soldier String Quartet has been fusing and confusing the notions of chamber and popular music. The SSQ appeared on records with John Cale, Robert Dick, Nicolas Collins, Phill Niblock, Elliott Sharp and Jonas Hellborg with tony Williams.

STATION ROSE
was founded in 1988 by Elisa Rose and Gary Danner; interface designers and propagandists for a new culture of conciousness who have their roots in the classic fine arts as well as the myths and rituals of pop culture. They have been producing recordings, videos, installations and fashion since 1979, and in 1991 Station Rose moved its base of operations to Frankfurt.

STADTWERKSTATT
Founded 1979. In the 80's, the primary concern was the creation of the center. The stage was freedom and freedom the stage. Dezentralization was imposed in the 90's which resulted in the creation of node points for a communicative sculpture. The stage is everywhere, it is the interface to dialogue.

CHARLY STEIGER
Born 1958. Ms. Steiger has been living in Frankfurt am Main since 1979. She has created spatial installations using various media (video, light) and musical projects in Frankfurt, Kassel, Cologne, Munich, Leipzig, Darmstadt, Mainz, Trier, Budapest, Perth, Linz, s'Hertogenbosch and Valencia.

GEROLD A. THALER
Born 1965 in Linz. Has studied since1987 under Prof. Gsöllpointner at the Academy of Artistic and Industrial Design, Linz (Master Class for Metal and Object Design). Works in metal and public sculpture, architectural conception, as well as product and furniture design.

MARK TRAYLE
1955 born in California. He attended the University of Oregon and Mills College, studying composition with Robert Ashley, David Behrman, and David Rosenboom. Performances in the U.S. and Europe. He received a grant from the National Endowment for the Arts and the City of San Diego. As a member of the computer network band, The Hub, he has performed at major festivals and venues throughout the United States and Europe.

JOACHIM SAUTER (ART+COM)
1959 geboren. Studium an der Hochschule der Künste Berlin und der Deutschen Film und Fernsehakademie, Berlin. Arbeit, Lehre, Veröffentlichungen, Austellungen, Preise im In- und Ausland. (u.a. Auszeichnung interaktive Kunst Ars Elektronica 1992 für „Zerseher"). Gründungsmitglied Art + Com, e.V. Berlin. Professor für das Gestalten mit digitalen Medien an die Hochschule der Künste Berlin.

ELLIOTT SHARP
Komponist/Multi-Instrumentalist, Kopf der Gruppen CARBON und ORCHESTRA CARBON, 'DYNERS CLUB, BOOTSTRAPPERS und TERRAPLANE. Zahlreiche Produktionen auf verschiedenen unabhängigen Labels in den USA, Kanada und Deutschland; neuere Produktionen mit CARBON, ORCHESTRA CARBON und 'DYNERS CLUB. DIGITAL ist auf der neuen Platte des Kronos Quartet enthalten. 1977 gründete Sharp das Label zOaR zur Produktion seiner eigenen und anderer Extremkompositionen.

JEFFREY SHAW
1944 in Melbourne geboren. Derzeit Direktor des ZKM Instituts für visuelle Medien, Karlsruhe. Ausgewählte Arbeiten: Corpocinema, „Sigma Projects", Museumsplein, Amsterdam 1967; Diadrama, Lantaren Theater, Rotterdam1974; Viewpoint, 6th Biennale de Paris, Musée d'Art Moderne, Paris 1975; Eve, „MultiMediale 3", Zentrum für Kunst und Medien (ZKM) Karlsuhe.

SOLDIER STRING QUARTET
Seit seiner Gründung 1985 beschäftigt sich das SSQ mit der Verschmelzung und Vermengung von Kammermusik und populärer Musik. Er hat Platten mit John Cale, Robert Dick, Nicolas Collins, Phill Niblock, Elliott Sharp und Jonas Hellborg mit Tony Williams aufgenommen.

STATION ROSE
wurde 1988 von Elisa Rose und Gary Danner gegründet, welche sich als Propagandisten einer neuen Bewußtseinskultur, als INTERFACE-DESIGNER verstehen, mit Roots in der klassischen bildenden Kunst, sowie in den Mythen und Ritualen der Pop-Kultur. Seit 1979 produzieren sie Tonträger, Videos, Installationen und Mode. Seit 1991 lebt und arbeitet Station Rose in Frankfurt/M.

STADTWERKSTATT
Gründung 1979. In den 80er Jahren stand das Schaffen von Zentren im Vordergrund. Die Bühne war der Freiraum und der Freiraum die Bühne. In den 90ern drängen sich Dezentralen auf, das bedeutet das Schaffen von Knotenpunkten einer kommunikativen Skulptur. Die Bühne erscheint überall, sie ist Schnittstelle zum Dialog.

CHARLY STEIGER
1958 geboren. Lebt und arbeitet seit 1989 in Frankfurt/M. Raumbezogene Installationen in verschiedenen Medien (Video, Licht) sowie Musikprojekte in Frankfurt/M., Kassel, Köln, München, Leipzig, Darmstadt, Mainz, Trier, Budapest, Perth, Linz, 's-Hertogenbosch und Valencia.

GEROLD A. THALER
1965 in Linz geboren. Seit 1987 Studium bei Prof. Gsöllpointner an der Hochschule für künstlerische und industrielle Gestaltung in Linz (Meisterklasse Metall und Objektgestaltung). Beschäftigung mit Metallplastik, Großplastik im öffentlichen Raum, architektonischer Konzeption, sowie Produkt- und Möbeldesign.

MARK TRAYLE
1955 in Kalifornien geboren. Studium an der Universität von Oregon und dem Mills College. Kompositionsstudium bei Robert Ashley, David Behrman und David Rosenboom. Auftritte in den USA und in Europa. Stipendium des National Endowment for the Arts and the City of San Diego.

WOLFGANG TSCHAPELLER

geboren in Dölsach, Osttirol. Tischlerlehre, Studium an der Hochschule f. angewandte Kunst Wien, Postgraduate Studies (MA) USA, Lehrtätigkeit, Cornell University USA. Arbeitet in Wien. Projekte (Auswahl): Trigonmuseum Graz, Berggasse Wien, Berlin Alexanderplatz.

WOODY VASULKA

geboren in Brünn, CSFR. Studium der Metall-Technologie und Hydraulischen Mechanik. Zusammen mit Steina hat Woody das New Yorker Medienzentrum The Kitchen gegründet. Er hat an vielen Ausstellungen in den Staaten und Europa teilgenommen, hat Vorträge und Vorlesungen gehalten und Artikel veröffentlicht, Musik komponiert und zahlreiche Videobänder hergestellt. 1979 erhielt er das Guggenheim Stipendium. Er lebt derzeit in Santa Fe, NM.

PETER WEIBEL

1945 in Odessa geboren. Studium der Literatur, Medizin, Logik und Philosophie in Paris und Wien. Lehrtätigkeit an der Hochschule für angewandte Kunst in Wien, am College of Art and Design in Halifax/Kanada, Gesamthochschule Kassel, Center for Media Study der University of New York in Buffalo. Seit 1989 leitet er als Direktor das Insitut für Neue Medien an der Städelschule Frankfurt. Zahlreiche Publikationen.

HANS PETER WÖRNDL

geboren in Salzburg. Absolvent der HTL /Hochbau. Studium der Architektur an der TU München (Kurrent), Postgraduate Studies und Lehrtätigkeit, Cornell University, USA. Arbeiten (Auswahl): Wohnbau Wittgensteingründe Wien (mit Rieder und Tschapeller), Wettbewerb Messepalast, Preisträger 1. Stufe (mit Rieder), Wiener Hochhausstudie (mit Rieder, Coop Himmelblau, u.a.).

WOLFGANG TSCHAPELLER

Born in Dölsach, Osttirol. Cabinetmaker, studied at the Academy of Applied Arts in Vienna, recieved his MA in the US and has taught there at Cornell University. Works in Vienna. A selection of his projects include the Trigon Museum in Graz, Berggasse in Vienna and Alexanderplatz in Berlin.

WOODY VASULKA

Born in Brno, CSFR. Studied metal technologies and hydraulic mechanics. Together with Steina he founded The Kitchen, a NYC media theater, and has participated in many major shows in the US and abroad, given lectures, published articles, composed music and made numerous video tapes. He is a 1979 Guggenheim Fellow and currently resides in Santa Fe, NM.

PETER WEIBEL

was born in 1945 in Odessa. He studied literature, medicine, logic, and philosophy in Paris and Vienna. He has taught at the Academy of Applied Arts in Vienna; the College of Art and Design in Halifax, Canada; the Gesamthochschule Kassel; and the Center for Media Study, University of New York in Buffalo. Since 1989 he has been Director of the Institute for New Media at the Städelschule, Frankfurt. Countles publications.

HANS PETER WÖRNDL

Born in Salzburg. Graduated from the technical school in Hochbau. Studied architecture at the TU Munich with postgraduate work at Cornell where he has also taught. A selection of his works include the Wittgensteingründe apartment complex in Vienna (together with Rieder and Tschapeller), an award of the first class for the Messepalast contest (together with Rieder), and the Viennese High-Rise Study (with Rieder, Coop Himmelblau and others).